THE
BIRMINGHAM GUN TRADE

THE
BIRMINGHAM
GUN TRADE

DAVID WILLIAMS

First published 2004
This edition published 2009

Reprinted 2010, 2011, 2016

The History Press
The Mill, Brimscombe Port
Stroud, Gloucestershire, GL5 2QG
www.thehistorypress.co.uk

© David Williams, 2004, 2011, 2016

The right of David Williams to be identified as the Author
of this work has been asserted in accordance with the
Copyrights, Designs and Patents Act 1988.

All rights reserved. No part of this book may be reprinted
or reproduced or utilised in any form or by any electronic,
mechanical or other means, now known or hereafter invented,
including photocopying and recording, or in any information
storage or retrieval system, without the permission in writing
from the Publishers.

British Library Cataloguing in Publication Data.
A catalogue record for this book is available from the British Library.

ISBN 978 0 7524 3237 3

Typesetting and origination by The History Press
Printed in Great Britain

Contents

	Acknowledgements	7
	Preface	9
1	An Introduction	11
2	The Manufacturing of the Brown Bess Military Flintlock	21
3	Armory Practice or the American System of Manufactures	55
4	England Embraces the American System of Manufactures	73
5	The Birmingham Trade and the British Machine Makers Respond	95
6	The Birmingham Trade from 1880	137
	Conclusions	167
	Endnotes	173
	References	183
	Index	189

For Ivy and Sid

'"Engineers whose opinions were worth having were too busy to write books", but now…'

(General MacNeill in the discussion following Colts' lecture to the Institution of Civil Engineers on revolver making in 1851)

ACKNOWLEDGEMENTS

To Sid Cruxton for starting off my interest in guns and gun making in the 1960s and restarting it in the 1990s, for sharing insights in the practices of the trade and for collecting many of the items photographed for the book from the trade as it contracted in the 1960s; to Professor Bill Johnson for his continued encouragement to scholarship, for sharing his work on Robins and the history of ballistics and for pointing me to the work of W.R. Moore; to Professor Paul Wright of the University of California Berkeley for continued friendship, for the creation of model texts in manufacturing and for pointing me to the work of Rosenburg; to Dr Jim Andrew of Thinktank, Birmingham, who helped source many of the illustrations, encouraged and critiqued the text and supported my search for a publisher; to Duncan Bedford for his help on early rifling technologies and for letting me have his *The Modern Gunsmith* by Howe; to Peter Jacques for help on rifling and the use of the Enfield gauges and for the loan of his excellent photographs; to the Muzzle Loaders Association of Great Britain for permission to reproduce many illustrations from its Journal and Newsletter *Black Powder*; to Bill Harding at the Birmingham Proof House for his help with illustrations, the Tower in Bagot Street, the Carr Brothers and many other leads; to Bill Curtis for his comments on the developing manuscript, for help with illustrations and Artifex and Optifex, and for emphasising the importance of the report of the Committee of 1854; to Peter Smithurst for his photographs of the Robbins & Lawrence milling machine; to Tony Murray for sharing his knowledge on lock making and detailed review of this manuscript; to Keith Moore of the Institution of Mechanical Engineers for identifying relevant material on Whitworth and the Birmingham gun trade in the Institutions archives; to Robert Gordon of Yale University for sharing his work on the real capabilities of the early American arms makers; to Peter Schmidt for his help on Hall Rifles; to Pat Malone for his encouragement and discussions on design for manufacture and firearms production; to Graham Greener for permission to reproduce the photograph of barrel welding at Greeners; to Hadyn Hill for allowing me to photograph his workshop; to Wayne Moore for permission to reproduce the figure of the Polhem gauge; to Dru Bronson-Geoffrey of the Springfield Armory for her help in locating the photograph of the Blanchard Lathe; to Paul Chambers for allowing me to reproduce the picture of his father Tom using the stock or draw knife; to Simon Clode for allowing me to reproduce the photographs of Westley Richards; to Richard and Molly Milner for locating the photograph of revolver finishing and Richard Jones of the MOD pattern room for permission to publish it; to Jim Reading for finding many of the excellent early gun books and encouraging me to write this work up; to Vince Scothern and Mary Treasure for preparing the photographs and tables, to Tony Woolrich, Andrew Lewer and Campbell McCutcheon for their help in moving the work to publication; and to Adrian Dent and Nick Fox for their friendship and a common interest in manufacturing techniques that were invented by the Victorians that are both difficult to replicate and allow us to learn about the future. To Ivy Kay Williams, Mum, for, among many other things, making me think about the relationship between the artisan and handicraft tradition of the old Birmingham trades and the newer engineering approaches.

I gratefully acknowledge the assistance received from Birmingham Museums and Art Gallery in tracing, and providing access to images of the gun trade used in this publication. The photographs shown as Source: Birmingham Proof House Library are reproduced by kind permission of the Guardians of the Proof House. The photographs of the workshops of G. M. Cruxton on pages 156-7 are held within the family; attempts to locate the copyright owner have failed as have attempts to locate the copyright holder of the maps from Smith. The images on pages 156-7 formed part of an oral history project and are also shown in White H. and Trudgeon R., Birmingham's gun quarter: A skilled trade in decline, Oral History, 11 (2), 1983, pp. 69-83.

PREFACE

Why write a book like this?

It all goes back to my parents: she was from Handsworth in Birmingham, he from Smethwick in the Black Country. Adjacent boroughs, but worlds apart in accent, pace and approach, both seminal places in the Industrial Revolution. The differences between them were deeper: she was immersed in the craft skills of the Birmingham metal trades; my maternal grandfather was a jeweller, a diamond setter and her brothers artisans, even artists, in stone, wood and metal and with the brush. He, on the other hand, was a machine maker, a designer of milling and grinding machines working for an American company, Cincinnati Milacron, and who had spent time working in the Soho Foundry while it was Avery's. His father had been a foreman, a toolmaker by training. These were mechanical engineers of the most traditional kind.

I followed Dad into engineering. For ten years from my mid-thirties as an academic I focused on the technology of electronics manufacturing and the way that the location of electronics manufacturing moved around the world as the business drivers and manufacturing technologies changed. I saw the battle between the manual processes of assembly using the dexterity of many people and the clumsy but tireless machine – the dextrous people with their low capitalisation seemed to win. I then moved to industry, and was responsible for manufacturing technology in a medical device company, making mechanisms in their millions out of plastic by machine. I had to think hard about precise interchangeable manufacture of safety critical mechanisms, to define the processes we used and to consider the automated solutions to be created.

Then I thought: 'I've heard about all this before, the volume manufacture of mechanisms, the battle between craft skills – handicraft – and the use of machines – the milling machine – and most significant of all the effect of changing technologies on people's jobs, livelihoods and locations.' I thought I should revisit and retell this story in order to learn from it in my work, I also had the time and inclination to do it as I was living away from home. That is why I wrote this book – I needed to resolve the role of those two cultures.

While finishing the book I returned to academia to continue my work in healthcare; when I returned I found that the concept of the economic cluster as identified by Michael Porter of Harvard Business School in his book *The Competitive Advantage of Nations* was as alive in healthcare as it was in electronics when I had left that sector four years earlier. An intent of writing this book is also to show that clusters were not invented in the 1980s, they have been around for over 200 years in some of the oldest manufacturing locations – and have always been disequilibrated by disruptive technologies!

This book is written by an engineer, not an historian, social historian or economist. Any errors of approach and detail are my own.

David Williams
Hathern, Leicestershire, 2003

1

AN INTRODUCTION

This book explores the relationships between the technology and history of gun making. It particularly focuses on the interaction between the methods required to achieve interchangeability in military firearms manufacture and the growth and decline of the Birmingham military and sporting gun trade. While this story has been told before, this work describes the production technology with more detail than is usual – 'technology and innovation is the historian's and economist's black box... the contents of the black box are difficult to unravel and their structure is difficult to perceive'.[1] The book captures the effect of these technical changes – one of them, the powered machine tool, a powerful disruptive technology – on an industry cluster in one of the world's oldest manufacturing locations.

'Interchangeability is the ability to interchange one of the components in an assembly, particularly a machine or mechanism'[2] without the requirement for adjustment or 'in making the parts of a mechanism so uniform in size that each mating part will go together and function properly without fitting'.[3] We can distinguish between practical interchangeability i.e. within a tolerance range or within limits and absolute or theoretical interchangeability. Interchangeability is of value in a military firearm for field repair – this was particularly realised in the early years of the United States – and for consistent performance, and has the additional benefit of allowing repetitive manufacturing at lower cost by the design of an appropriate manufacturing system. It is particularly important in very high – mass – production volume manufacturing. Interchangeable manufacture is usually achieved via the factory system. This is in direct contrast to a largely hand crafted 'best gun', which is individually tailored to a customer usually within a trade that is still largely based upon the domestic system of manufacturing. Joseph Roe, in his book *English and American Tool Builders* written in 1916, was the first to begin seriously exploring the subtleties of this journey from craft to engineering discipline.

In this chapter we outline the technology of the firearm and the place of Birmingham in the Industrial Revolution before expanding on the main theme: the Birmingham gun trade and the impact of technology change on it.

Firearms Technology

During the period of interest of this work, from around 1720 to 1950, the mechanical engineering technology of the 'lock, stock and barrel' firearm changed significantly. It began the period as a weapon with a flintlock with the demanding requirements of correctness of form and spring stiffness and setting of the variable flint rather than mechanism precision, through the deforming projectile technology of the lead Minie bullet muzzle-loading rifle with its simplified percussion lock that had fewer and more reliable components – the deforming bullet accommodating some geometric variation particularly in the barrel bore, to the unforgiving geometric tolerances of the magazine breech loader and its metallic cartridge. **1.1** and **1.2** show characteristic weapons of each era. At the top of the figures we have an India Pattern Flintlock Brown Bess of the late eighteenth century and below this an 1853 Pattern Enfield percussion Minie rifle (the first British weapon to be machine built to interchangeable principles and the basis of the Snider conversion to breech loading). The next weapon is a Snider breech loading conversion of the 1853 Enfield, the first widely adopted British breech loader. Below this is a Martini Henry breech loading rifle: the Martini Henry was adopted in 1871 as the first purpose designed single shot British military breech loader. With this increase in functional requirements – a more precise mechanism – came the growth of increasing manufacturing precision and its partner, measuring precision. If we also examine the evolving designs we see the artistic curves of the flintlock using a leaf spring as its energy store changing to the engineered designs that are easier to create by machine, including the flat surfaces of the breech loader and its coil spring energy store. The single shot cartridge breech loader was subsequently replaced with bolt action magazine rifles – the Lee Metford adopted in 1888 and Lee Enfield being the most famous British (British American) variants. The lowest weapon in **1.1** and **1.2** is a No.7 variant of the Lee Enfield. Key weapons from the USA in this period are the Hall Model 1817 breech loading rifle, the Colt revolver and the Spencer repeating rim fire cartridge carbine.

1.3 shows the items associated with the Enfield rifle, a socket bayonet and its scabbard, a pack of paper cartridges – each cartridge of greased paper contains a lead Minie projectile and black powder, and a multipurpose tool. The multipurpose tool includes

AN INTRODUCTION

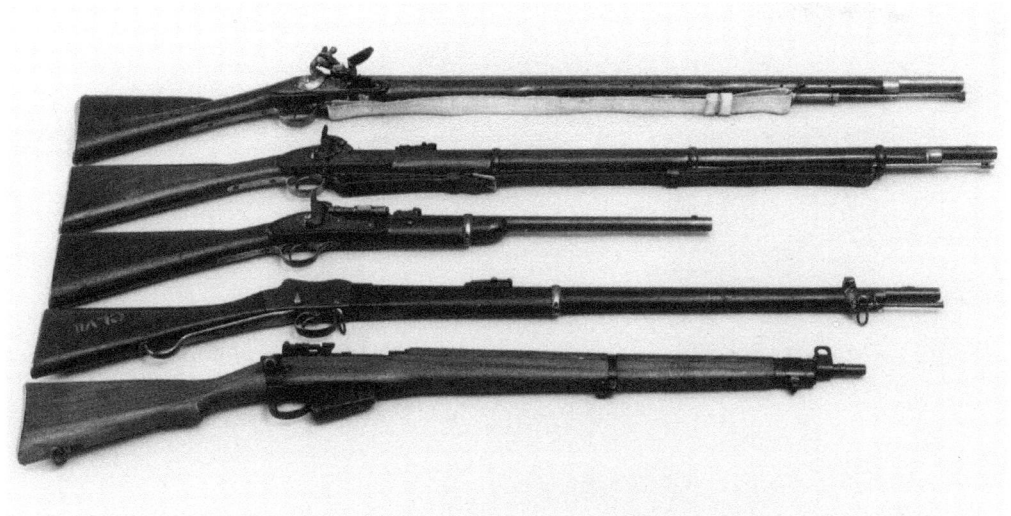

1.1 (Above) *Typical military firearms from 1750–1950.*

1.2 (Right) *Actions of military firearms, 1750–1950.*

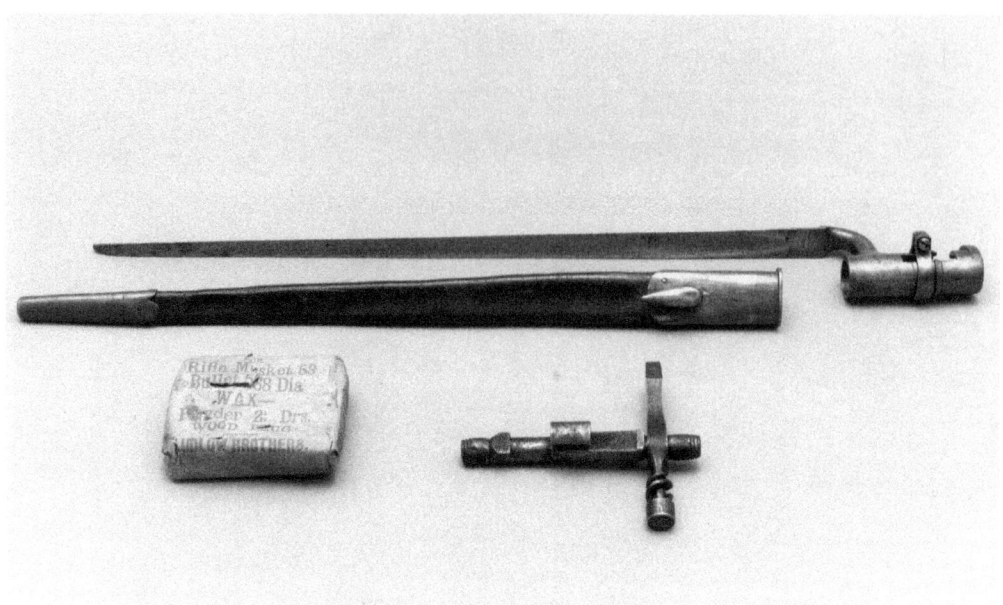

1.3 *An Enfield socket bayonet, pack of paper cartridges and Enfield tool.*

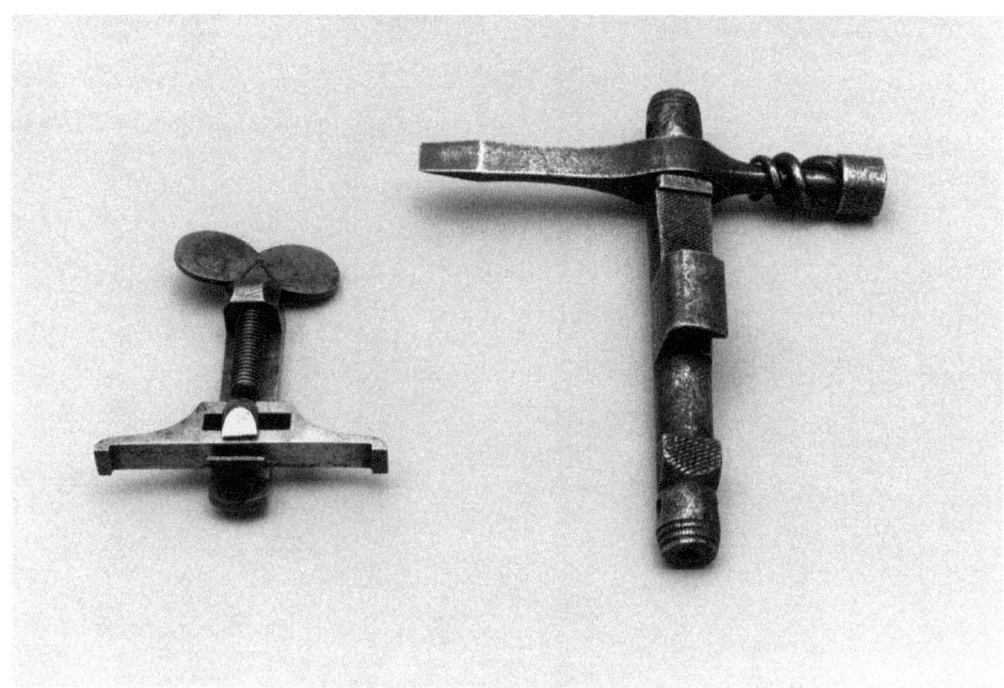

1.4 *Contrasting an Enfield tool with a sporting gun spring vice.*

AN INTRODUCTION

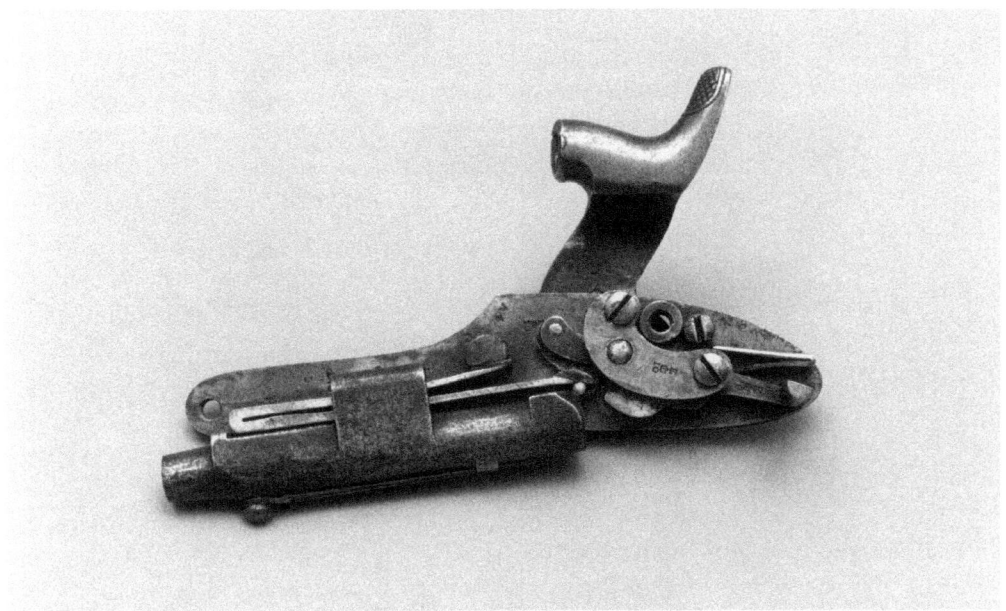

1.5 *Using the Enfield tool to compress the mainspring of an Enfield lock.*

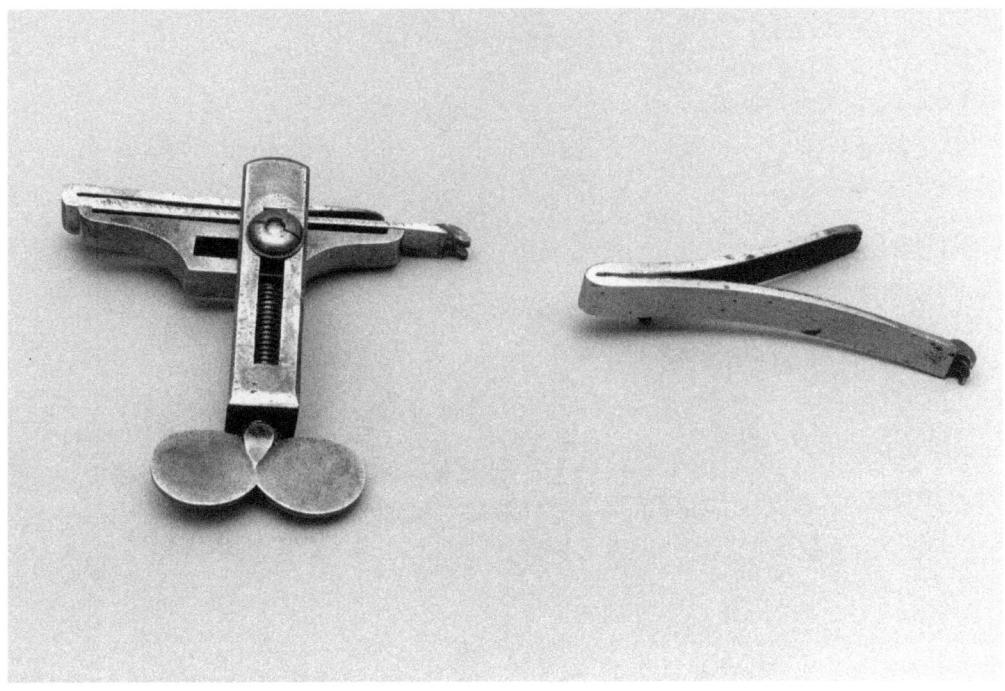

1.6 *A free Enfield spring and one held in a spring vice.*

THE BIRMINGHAM GUN TRADE

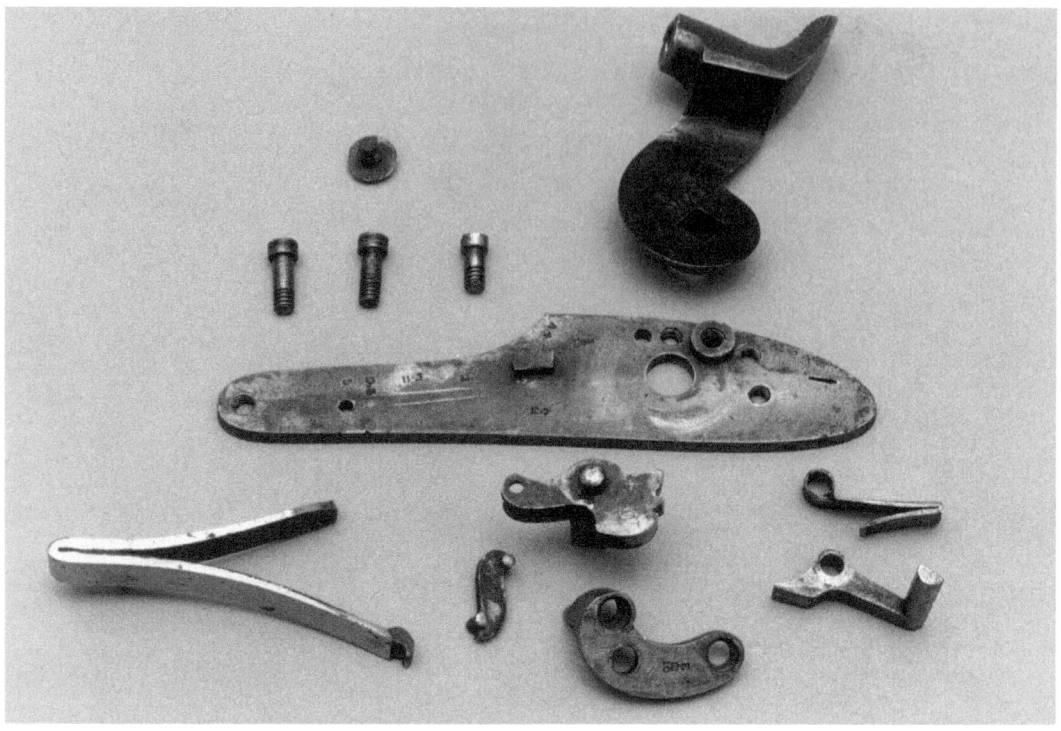

1.7 *The disassembled Enfield lock.*

an oil bottle, nipple pricker, screwdriver, worm, spike and mainspring vice. **1.4** shows the robust Enfield tool and a well-finished sporting gun mainspring vice. In **1.5** the Enfield tool is shown compressing the mainspring of a detached Enfield lock at half cock, the safe position, as the first process in disassembling a lock. When the spring is removed it can be held compressed in the vice – it is likely to fail if rapidly decompressed. The contrast between compressed and uncompressed springs is shown in **1.6**. The components of a lock are shown in **1.7**. The photograph shows the hammer or cock, the bridle, the fixing pins (screws) and the lock plate. The components below the lock plate are the main (left) and sear (right) leaf springs and the tumbler and swivel or link. This lock was made at the Enfield factory in 1864. The small marks on the lock components are made by inspectors stamps. **1.8** shows the contrast in the design and construction of the later Martini action. The excellent book *Live Firing of Powder and Ball Small Arms* by Martin Pegler,[4] shows the techniques used in shooting such weapons. This work builds strongly on key sources in the literature.[5]

AN INTRODUCTION

The Illustrations

The illustrations included in the work, particularly the photographs, have been chosen as much as possible to show the original machines and techniques used. Many of the photographs of the manual methods were taken in the original location of the trade in St Mary's, Birmingham. They therefore, although being relatively modern being taken in the 1950s, 1960s and 1970s, show the traditional hand techniques of the trade being carried out in their traditional surroundings in eighteenth- and nineteenth-century buildings. We should recognise that many of the hand techniques used then and now in the best gun trade were equivalent to those used in the military trade 150 years earlier in many of the same buildings.

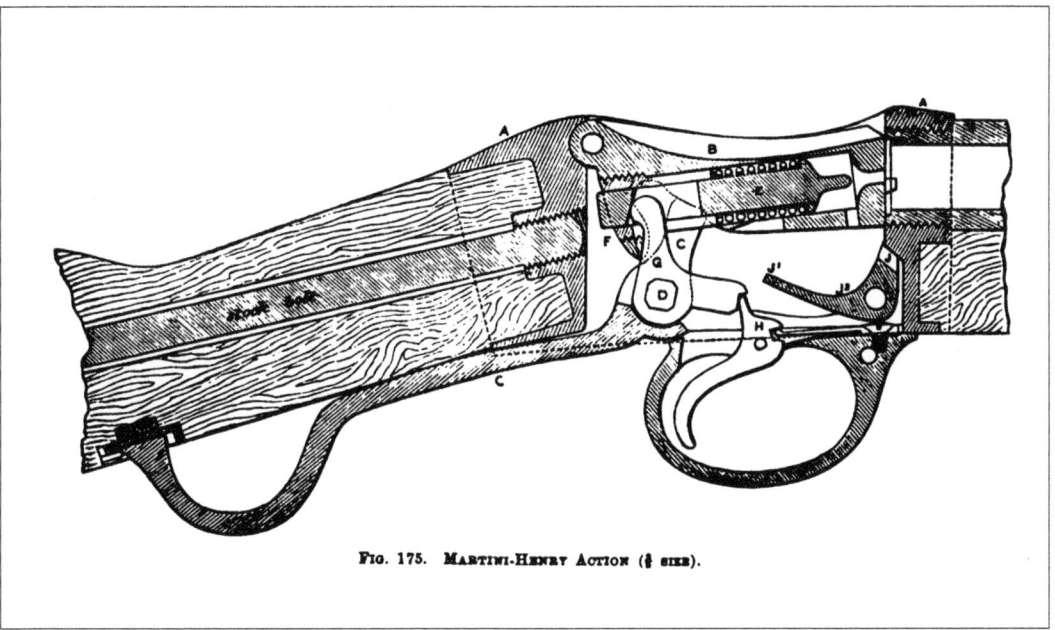

1.8 *Section of the Martini Henry action (Walsh 1884).*

Industrialisation and the English Midlands

The Industrial Revolution in the English Midlands may be considered to have begun when Abraham Darby developed the *volume* processes using mineral fuel to create pig iron in Shropshire in 1709. The understanding of this process and the location of rich factors of production in the west of the Midlands (iron ore co-located with coal and wood) led to the growth of the iron industry in the Black Country during the late eighteenth and through the nineteenth century. The extraction of the mineral fuels was assisted by pumps driven by the steam engine, the first Newcomen engine to be supplied to a coal mine was installed near Dudley Castle in 1712.[6] The Black Country, essentially South Staffordshire, was *black by day and red by night* being lit by its furnaces. This growth was of course assisted by the further technological innovations of James Watt working in partnership in Birmingham with Matthew Boulton from 1775, and John Wilkinson, an iron master from Shropshire. Together[7] they created more efficient engines for blowing the furnaces and driving the forges and rolling mills.[8]

Wilkinson's primary contribution was to develop a boring machine, perhaps the first modern machine tool[9] capable of accurately generating steam engine cylinders: 'there was no one with the skill or the means to bore a true cylinder 6in in diameter by 24in in length.' Previous best practice is captured by: 'a true cylinder bored on Smeaton's machine was ⅜in out of truth'.[10] The commercialisation of the steam engine is perhaps the most dramatic example of the synergy of the availability of raw materials, manufacturing methods and original science with the market demand and energy of an entrepreneur.

In 1775 Wilkinson developed his machine, basing it upon the cannon borer that had been used to bore Watt's first cylinder. It generated a boring by rotating the cylinder around a fixed axis with a cutting tool supported on a stiff boring bar passing through the whole cylinder and supported at each end. Matthew Boulton wrote in the following year: 'Mr Wilkinson has bored us several cylinders almost without error; that of 50in diameter that we have put up at Tipton does not err to the thickness of an old shilling in any part'. As we have said, this technology built upon that developed for cannon boring and of the Coalbrookdale Co. and Carron Co. in the machine finishing of large castings. The work of Jan and Peter Verbruggen, Masters of the Royal Brass Foundry at Woolwich,[11] shows exquisite watercolours of the early processes used to cast and bore cannon in the eighteenth century. Woolwich was an important site for the creation of the next generation of machine makers; Henry Maudslay worked in the Arsenal there as a twelve-year-old. Maudslay's subsequent pupils included Joseph Whitworth, James Naysmth, Richard Roberts (the inventor of the planing machine) and Joseph Clements.[12]

Birmingham has been a centre for the metal industries from the time of Henry VIII. 'A great part of the town is maintained by smiths who have their iron and sea cole out of Staffordshire' and through the sixteenth century 'full of inhabitants

AN INTRODUCTION

and resounding with hammers and anvils, for the most of them are Smiths'.[13] The 'workshop of the world' and 'the city of a thousand trades' was industrialised by the energy of such entrepreneurs as the Birmingham-born Boulton. He started to build the Soho Manufactory in 1761 – one of the first large manufacturing enterprises – in open country between Birmingham and the Black Country (near Handsworth, Birmingham).[14] Boulton also applied steam-driven presses to the coining of large copper 'cartwheel' one and two penny pieces in the Soho Mint, these being circulated in 1797. The Midland entrepreneurs had an interest in technology and scientific development, perhaps the most significant evidence of this being the Lunar Society – so called because it met on moonlit nights to make travelling easier – whose membership included, for example, the scientists Watt and Priestley and the entrepreneurs Wedgewood and Boulton.

Gun making was one of two of the earliest metal or 'hardware' trades of Birmingham; the other was the metal 'toy' trade.[15] These craft-based trades established Birmingham's reputation as the 'other face'[16] of the Industrial Revolution. One of the 'other face' characteristics of the industrialisation of Birmingham was the small and bridgeable gap between artisan and master – leading to a culture of respectability, self improvement, independence and sometimes an unwillingness to confront the master.[17] However, Joseph Tucker said in 1757 that Birmingham products were 'so exceedingly cheap to astonish all Europe' and that 'almost every Master Manufacturer hath a new invention of his own, and is daily improving on those of others'.[18] The skills learned in these trades were important and transferred to other trades in the same location. Boulton wrote to Smeaton in 1778:

> As we have done before in the button manufactury; we are training up workmen and making tools and machines to form the different parts of Mr Watt's engines with more accuracy, and at a cheaper rate than can possibly be done by the ordinary methods of working.[19]

Four chronological stages of industrial development have been identified in the West Midlands:[20] the long preparation for the fundamental changes that we call the Industrial Revolution; the core of the Industrial Revolution in the eighteenth century; the period of industrial prosperity during the first half of the nineteenth century that grew directly from the revolution; and what the Birmingham economic historian G.C. Allen called the 'New Era' from 1870,[21] which prepared the region for its twentieth-century industrial character. The engineer William Fairburn, in his Presidential inaugural address to the British Association in 1861, said that when he first went to Birmingham in 1814 all machinery was made by hand but that now 'everything is done by machine tools with a degree of accuracy which the unaided hand could never accomplish'.[22] This book follows these four phases for one important industrial sector, focussing on the reasons for and impacts of, the transition that Fairburn saw.

Structure of the Book

This chapter has briefly reviewed firearms technology and summarised the role of Birmingham in the early Industrial Revolution. The next chapter presents the processes for manufacture of military firearms in Birmingham before the 1850s – these being largely those still emphasised in the manufacture and finishing of high quality sporting guns – and describes the growth of the firearms cluster in the St Mary's area of Birmingham. Chapter Three turns to focus on the development of the techniques usually understood as 'Armory Practice' or 'The American System of Manufactures', primarily the use of special purpose machines for wood and metal working and the use of sets of precision – receiving – gauges. In Chapter Four we examine the reaction to this technology in the UK as promoted by Colt and others at the time of the Great Exhibition and installed in the Royal factory at Enfield in the 1850s. We then turn to examine the reaction of the Birmingham gun trade to this and the formation of Birmingham Small Arms and other companies using machines in Chapter Five. This chapter also shows that UK machine makers such as Thomas Greenwood of Leeds and James Archdale of Birmingham learnt much from imported technology. The techniques of rifling and cartridge manufacturing are also presented. Chapter Six follows the decline of the trade from the 1880s through the twentieth century to today and includes a number of excellent photographs of the trade taken by the Birmingham Museum of Science and Technology in the 1950s and 1960s. Through the book we see the steady increase of precision of manufacturing over time in gun making, the key technology that allows interchangeable manufacture and the evolution of breech loading and cartridge weapons and the transfer of these techniques to other products.

2

THE MANUFACTURING OF THE BROWN BESS MILITARY FLINTLOCK

One of the dominant Birmingham metal trades was the making of guns for the military and the slave trade.[1] Volume military gun making had begun with a significant contract via Sir Richard Newdigate in 1689,[2] although there had been weapons supplied to the Parliamentarians in 1643 and to the Ordnance in the late 1670s.[3] There were then no guilds in Birmingham to protect trades by demanding that all those who practised a trade had served apprenticeships. This led to the important step of the highly specialised approach of the division of labour and specialisation of task or manufacturing process by the mid-eighteenth century – there were at least thirty different 'sub-trades' or manual manufacturing processes in gun making.[4] By 1707 there were 400 gun and gun lock makers and their families in Birmingham.[5] One of the earliest accounts of this division of labour comes from the Birmingham, Wolverhampton and Walsall Directory of 1767:

> Gun and pistol makers... of these branches it is necessary to observe the number of hands they go thro' before completed, viz. the barrel maker who wields the barrel, the borer, the filer, then 'tis proved, after this it goes to the ruff stocker; in the lock branch there is the forger and filer, and dependent on these are the furniture castor, the engraver, polisher and filer, who are the gun and pistol makers; it is needless to say anything further of this trade, but to observe from this place all the kingdom is supplied with barrels and locks and the consumption abroad is very great.

This chapter describes the techniques and evolution of the Birmingham trade.

The Ordnance System of Manufacture

The UK's approach to making large numbers of military firearms between the 1720s and the mid-nineteenth century was the Ordnance system of manufacture – the system being well established in Birmingham by the 1720s.[6] The Ordnance system is essentially the sub-contracting of manufacture from the national armoury at the Tower of London. There is a late account of this[7] by John Marshall, Clerk and Paymaster of the Ordnance in Birmingham in 1824. Components were made by many sub-contractors in small workshops and individual components and mechanisms such as locks were brought together to be 'set up' by craftsmen finishers, either at the Tower or at sub-contractors. Setting up was a series of four manufacturing and viewing (inspection) steps – including 'rough stocking', 'screwing together' and 'making off' – that put the weapons together. A 'pattern', a weapon correct in every detail, was used as a master and copied by all. Perhaps unsurprisingly, quality in these circumstances was variable, especially from overseas suppliers: for example in 1794, 10,000 stand of arms had been ordered from Liège but 'they cannot work to a pattern and although the bore is the same, scarce any two of the muskets are similar.'"

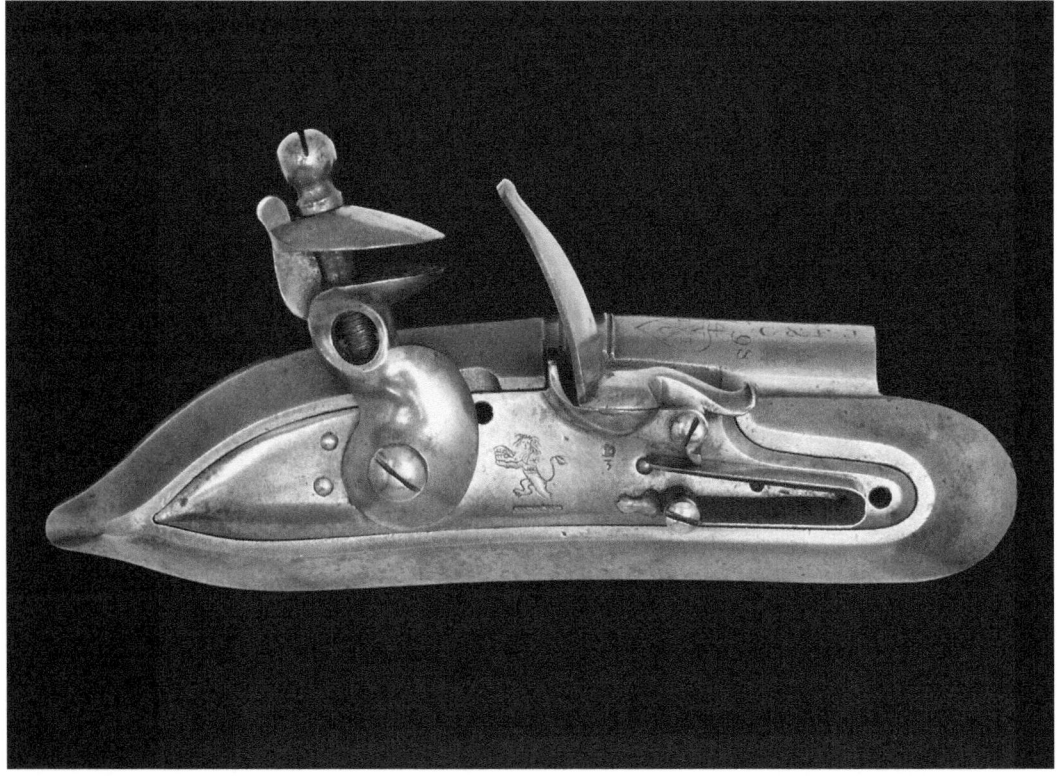

2.1 *An East India Co. jigger of 1798, engraved No.15. (By permission of Christies Images Ltd)*

Periodically, steps were taken to improve the standards of the Birmingham contractors to the Ordnance. In July 1716, Richard Wooldridge, Master Furbisher at the Tower, went to Birmingham 'to show ye workmen ye way to fitt locks to the mould'. Later in 1755, Richard Davis was sent to Birmingham 'to give the workmen proper directions and to see that their tumbler tools, screw plates and taps are all made to the same gauge'. A year later the locks were 'improved, better filed and all the parts made to particular gauges'. The East India Company had also used jigs to test fits since 1781. An East India Co. jigger of 1798 is shown in **2.1**.[8] This jigger is marked with 'No.15', indicating that it was one of a set of at least fifteen gauges. Government viewing rooms[9] in Bagot Street, Birmingham known as the 'Tower' were subsequently opened in 1797.[10] They closed in 1888[11] and were occupied by W.W. Greener in 1914.[12] **2.2** shows a plan of the Tower in its heyday. The government viewers had high standards and frequently rejected work; the gun makers therefore preferred to work for the East India Co., Hudson's Bay Co. and the Royal Africa Co. 'whose controls were laxer and payments quicker'.[13] In spite of this, Ezekiel Baker, writing in 1832 in an edition of his *Remarks on Rifled Guns* campaigning for tighter standards of proof, comments on Tower Musket barrels that the 'Inspector at Birmingham had

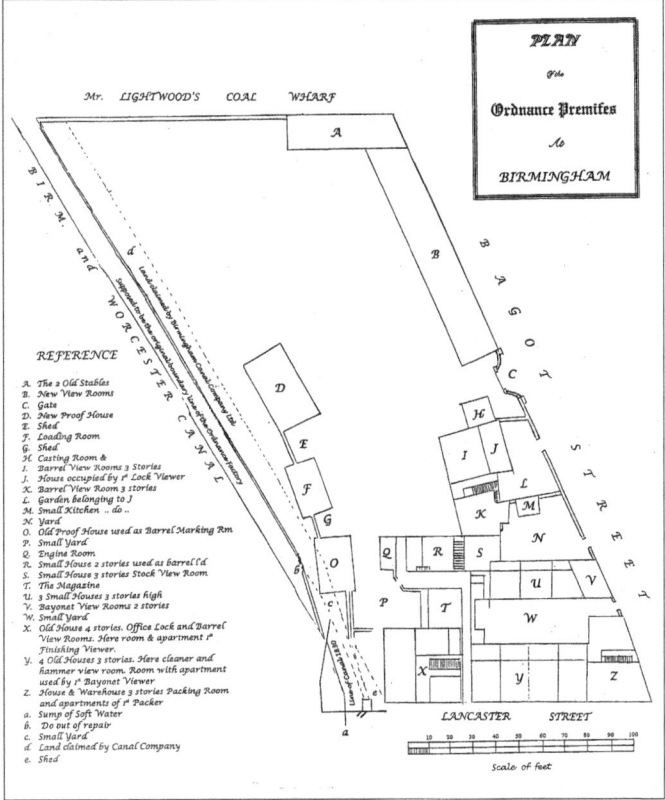

2.2 *Plan of the Tower, Bagot Street. Birmingham Proof House Library. (By permission of the Guardians of the Proof House)*

2.3 *A Birmingham workshop showing filing as the primary manufacturing process. An action body is being filed up to fit the ejector mechanism in the fore end of a gun. (By permission of the Birmingham Museums and Art Gallery)*

totally neglected his duty – indeed he seemed to be of the opinion that *quantity* not *quality* was the only desideratum' (his italics).

The Machines and Manufacturing Processes Used

The key manufacturing process being used is fitting – the careful hand finishing, usually by filing, of individual components such that they fit together, or to a gauge. **2.3** shows a gas-lit traditional hand workshop in Birmingham with a sporting gun mechanism being filed in a leg vice with lead jaw protectors. **2.4** shows some hand tools collected from the Birmingham trade: draw knives for stock making, gun maker's braces for driving screws; – 'nails' in gun making – and cutting tools, single and double barrel breech plug wrenches, a cast off stick (for fitting sporting gun stocks) and a trigger pull spring balance. **2.5** and **2.6** show hand tools: screwdrivers – known as turn screws – and wrenches, a simple bore gauge[14] and a screw plate for cutting screw threads and a sprung vice jaw pair for holding components as they are filed – known as a pin clamp

THE MANUFACTURING OF THE BROWN BESS MILITARY FLINTLOCK

by the trade. As Wilkinson, the noted London gun and sword maker notes in his book *Engines of War* on guns and gun making, published in 1841, the way to make locks is 'good filing and fitting with excellent tools'. Fitting, particularly of metal to wood, within the gun trade, has traditionally been achieved with the aid of the smoke lamp. A smoky flame from a paraffin lamp blackens the components to be fitted together, and then components are assembled. After disassembly, black spots on the wood show high spots to be removed, as bright spots are shown with Engineers Blue.

Much of the material used and many of the locks came from the nearby Black Country towns of Bilston, Darlaston, Wednesbury, Willenhall and Wolverhampton,[15] with perhaps the most significant manufacturers being the Brazier's of Wolverhampton. A listing that goes up to the present day identifies 400 individuals and firms in the lock making and gun barrel trades in Staffordshire, together with a number of gun and pistol makers.[16] Wilkinson's rival, William Greener, the Newcastle gun maker who moved to Birmingham to be at the centre of the trade, wrote about lock making in his book *The Science of Gunnery* on Brazier, in 1841. He said 'There is more real science displayed in the construction of a gun lock than mechanics in general imagine... The quality of all locks depends on the price they cost filing'. Braziers made locks for the Board of Ordnance and for the East India Co. Their works were known as

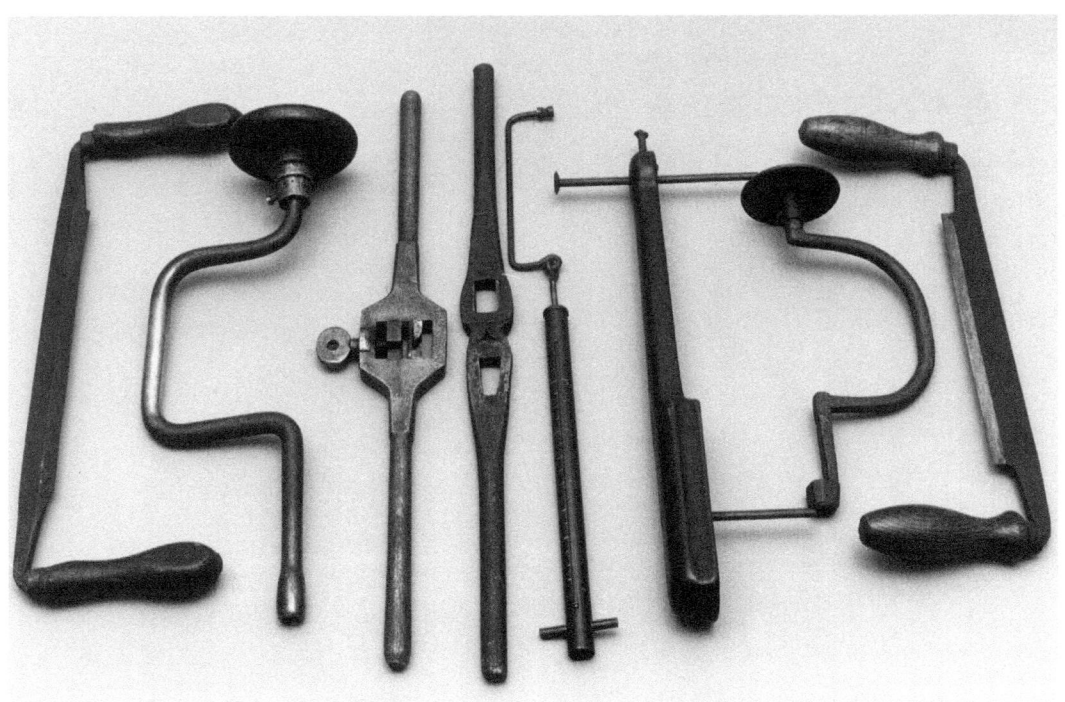

2.4 Hand tools from the Birmingham trade. Two stock knives, two gunsmith's braces, barrel plug wrenches for single and double barrelled muzzle loaders, a cast off stick and a trigger pull spring balance.

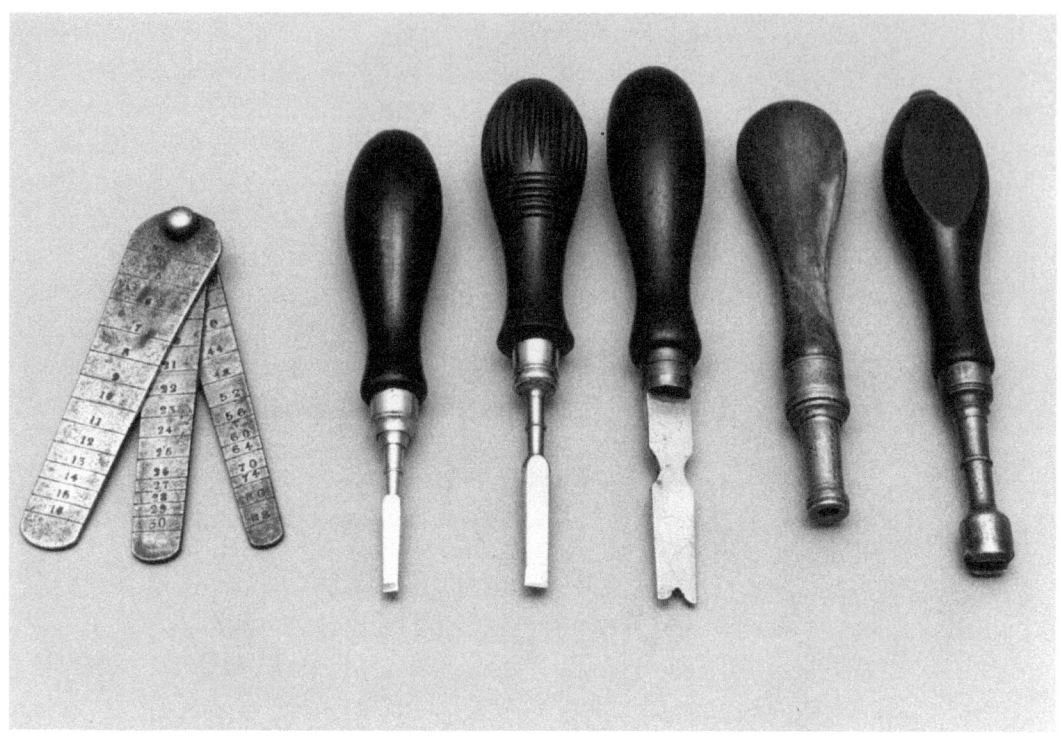

2.5 *A bore gauge, turn screws and wrenches.*

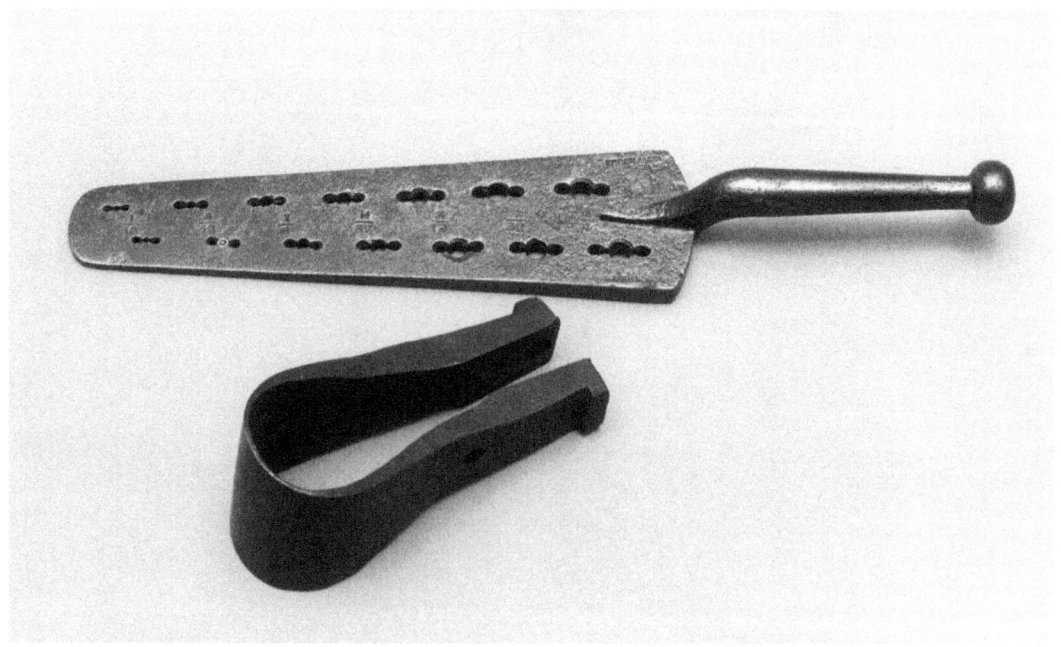

2.6 *A screw plate and pin clamp – sprung vice jaws for clamping components without damage.*

Ashes and were in Great Brick Kiln Lane, Wolverhampton.[17] Darlaston had 300–320 gun lock filers, 50–69 gun lock forgers and 250 boys employed as lock forgers and filers, cock stampers and pin-forgers in the mid-eighteenth century.[18] As we will see in subsequent evidence to the review of machinery in the manufacture of military firearms in the 1850s, Admiral Hastings notes that Braziers are considerable users of machinery in lock making.[19]

Bodmer, of whom we will hear more later, visited Birmingham in November 1816[20] as part of a tour to find out more about the manufacturing technologies applied in England. He comments on his second day in the city: 'The number and the variety of the factories in Birmingham is quite extraordinary. But there is more bad workmanship here than good workmanship'. Bodmer was trying to source both locks and barrels from the Birmingham trade. He visited a barrel maker who, with two assistants, made a barrel in 'rather under 45 minutes'. He spent a considerable time with Mr Little, perhaps of St Paul's Square[21] and asked him to make locks and a barrel to his pattern. Unfortunately for Mr Little, Bodmer writes of the locks: 'I do not think they were a great success' and of the barrel: 'it was not a good barrel and more faults became apparent when it was tested'. Bodmer returned to his manufactory in the Black Forest but following dissatisfaction with his employer and the death of his wife, became restless; he worked in the north-west of England between 1824 and 1829 and between 1833 and 1847, making many mechanical engineering innovations.

As we can see in the manual, trade forging was carried out;[22] hot stamps were noted to be used in the Birmingham button trade in 1749 by Samuel Schroeder, a visitor from Stockholm, in his day book.[23] Stamps are one of the earliest examples of the use of tools to achieve standardised products.[24] In 1816, Bodmer also notes the use of 300 and 400 pound drop hammers in a Birmingham wire factory.[25] When in January 1815, Price & Co. of Birmingham began stamping out cocks of flintlocks by machinery, three screw presses and a stamping machine were ordered for the government factory at Lewisham to duplicate the process.[26] In addition to the cutting processes mentioned further on, John Jones of Birmingham is said to have introduced a number of forging techniques between 1819 and 1825 when he moved to Russia to work in the Tula Armory, including techniques of barrel welding on the anvil and the use of drop stamps for lock component forging and fly presses for cropping the excess material from the forging.[27] In 1838, George Round, Lock Filer and Samuel Whitford, Die Sinker took out a patent for 'Manufacturing Gun Locks by Stamping'.[28] Deane, in his *Manual of Firearms* of 1858, notes that the Birmingham trade had used simple machinery for many years, including that for stamping and forging. As James Nasmyth, the inventor of the steam hammer, notes on die forging while discussing[29] his own use of moulds or dies to stamp masses of hot iron in his steam hammer in the 1840s: 'This practice had been in use on a small scale in the Birmingham gun trade. The iron-work of firearms was thus stamped to exact form'. Harpers Ferry in the United States

2.7 *Hammer forgings and stampings collected from the Birmingham gun trade.*

2.8 *Two hand- or hammer- forged hammers collected from the Birmingham gun trade.*

THE MANUFACTURING OF THE BROWN BESS MILITARY FLINTLOCK

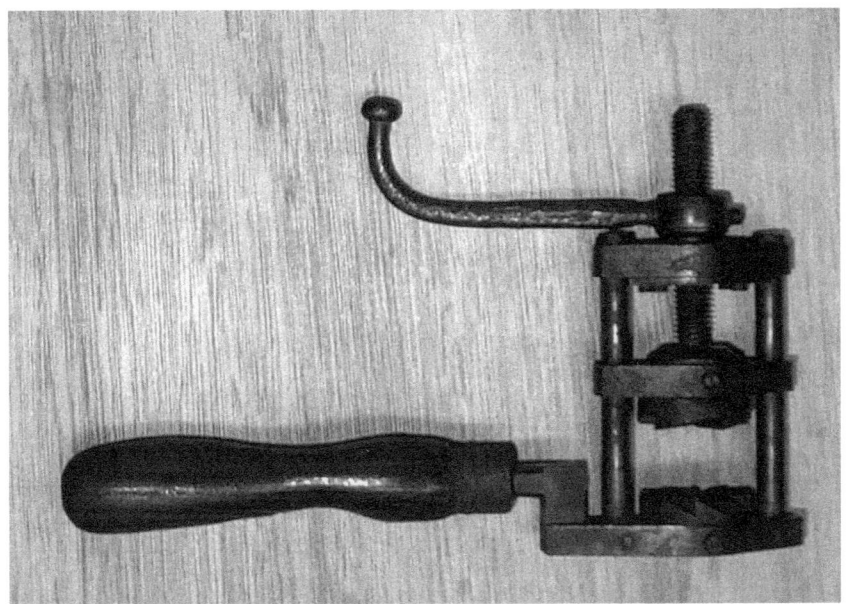

2.9 *A tumbler mill, a hand-held machine for machining the faces of a lock tumbler. (By permission of the Birmingham Museums and Art Gallery)*

were die forging in 1827.[30] W.W Greener in 1910 shows a variety of hand and machine percussion and centre fire hammer forgings,[31] see also those in **2.7**,[32] including flintlock cock forgings, and **2.8**, close ups of the hand or hammer forgings, collected from the Birmingham trade and elsewhere in the 1960s. Clearly the complexity of the forging process increases with the size of component being manufactured and the degree of precision required, the force required increasing with both. The best hammers were filed from hand forgings and stampings used for lower quality guns.[33]

More importantly, some small hand-held, hand-powered metal cutting machines were also used.[34] See for example a tumbler mill in **2.9** and a screw mill by Charles Whitehouse, small tool maker to the trade, in **2.10**. Gamel's description of the machinery used at Tula in 1826 shows a similar tumbler mill, 'an English-made device called the turning device', perhaps introduced by John Jones. The two types of mill have characteristic multi-point cutters and are hand- or vice-held machines which are either hand- or brace-driven and are used to finish rough forged components. They particularly ensure that the two surfaces of the tumbler are parallel and generate the shafts perpendicular to the surfaces by the process of hollow milling. That these are long established processes is confirmed by Diderot, showing tumbler mills based on files in gun making practice in his 1751–1765 encyclopaedia.[35] In the UK we have used processes similar to milling for many years; some say milling

2.10 (Above) *A screw mill by Whitehouse who occupied 13 Weaman Row between 1830-1850.*

2.11 (Left) *Bullet mould cherries.*

was invented by Robert Hooke (said in the discussion following Colt's lecture at the Institution of Civil Engineers, see later), and in 1829 Nasmyth built a special purpose hexagon nut milling machine as Maudslay's assistant. **2.11** shows Birmingham 'cherries', hand-cut ball cutters for the manufacture of bullet moulds.[36] **2.12** shows a bullet mould (incidentally for a belted or Brunswick ball) from around 1840, together with a rough casting for finishing into an arm of a mould.

More machinery was routinely used for barrel grinding and boring, perhaps because of the physical size of the components and the consequent power requirement and the bore accuracy required. This machinery was located near water, the source of power. Barrel makers were located in Aston, Deritend, Smethwick and West Bromwich, Smethwick and West Bromwich being on the Birmingham – Black Country boundary.[37] The next step was the use of steam power to drive the same machines. Ketland, one of the major Birmingham gun makers, in Whittal Street, used a steam engine in barrel making and grinding;[38] Whatley of Smethick in 1801 'has established a large manufactury of gun barrels which are forged and bored by the aid of a steam engine';[39] the New Steam Mill Co., barrel makers, was founded in Fazeley in 1803 with a 56hp engine and in 1811 Louise Simond visited a mill where 300 men produced 10,000 barrels a month using a 120hp engine.[40]

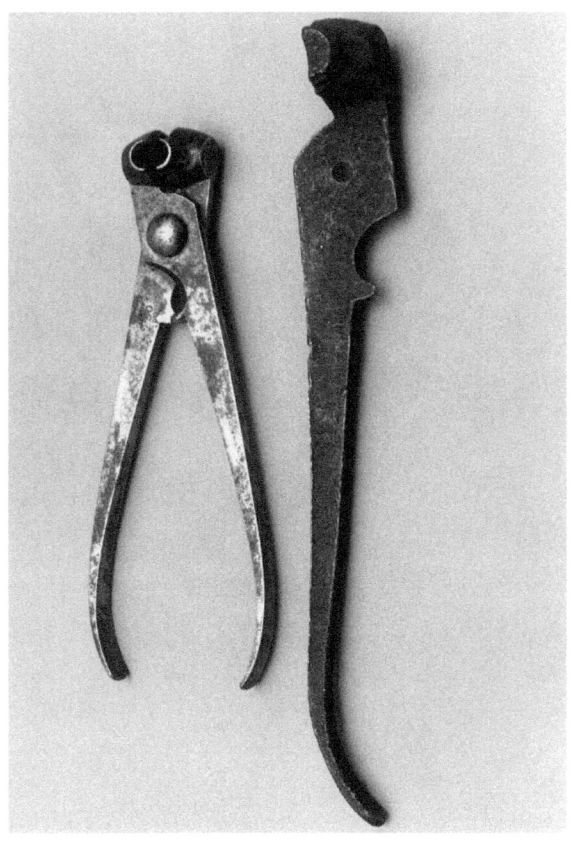

2.12 *A rough casting for a bullet mould component and a bullet mould from around 1840 for a Brunswick rifle.*

2.13 *Metal rolling, from W. Greener,* The Science of Gunnery as applied to the use and construction of firearms, *Longman, 1841.*

In 1814, a Prussian civil servant, Johann May, reported to his minister on a trip to England[41] to give us the following description of a barrel making factory:

> I visited a Birmingham gun factory in which various rolling mills, grinding mills and boring machines were operated by a 110hp steam engine. Many hundreds of workers are employed there. The factory is mainly engaged in fulfilling orders already received. The factory not only makes complete guns but it will also undertake rolling or grinding or boring of single pieces for its customers. The machines in this factory are extremely efficient. They work very quietly and there is little noise in the plant.

In his 1841 book, William Greener shows us woodcuts of metal rolling, **2.13**, and barrel boring, **2.14**, by machine and the manual process of twist barrel welding at the forge, **2.15**, and the coil being welded, **2.16**. Such Damascus barrels were widely used for shotguns through the eighteenth and nineteenth centuries and were made by hammer welding coils of specially prepared twisted rods, the twist and its materials giving rise to the figure of the barrel – see **2.17**. The coils were 'jumped' to weld each turn to the next and then completed by hammering

THE MANUFACTURING OF THE BROWN BESS MILITARY FLINTLOCK

2.14 *Barrel boring, from W. Greener, The Science of Gunnery as applied to the use and construction of firearms, Longman, 1841.*

3in-lengths at a time at welding heat in a U-shaped anvil. These barrels allowed the percussion shotgun to be very light and were forged close to their final dimensions so that the toughened skin produced by hammering was not filed away when the barrel was finished.[42] **2.18** shows the welding of a Damascus barrel by a gaffer and two workmen on a formed anvil next to the hearth at the Greener factory, very early in the twentieth century.[43] The workmen have large hammers and the gaffer a small hammer for the precise touches. Greener's were the last company to make Damascus barrels in England, as production ceased in 1903. Barrel grinding is shown in **2.19**.

> Grinding[44] is evidently the roughest and most laborious process connected with gun making, and in this work skill and great practical experience are essential. The ponderous grindstones used are from Derbyshire, and as the huge discs rotate sturdy workmen press the barrels against the stones, which flash forth a blaze of sparks, reducing the metal to the required size with nicety and exactitude, the gauge being constantly applied to ascertain results.

2.15 *Twist barrel welding at the forge, from W. Greener,* The Science of Gunnery as applied to the use and construction of firearms, *Longman, 1841.*

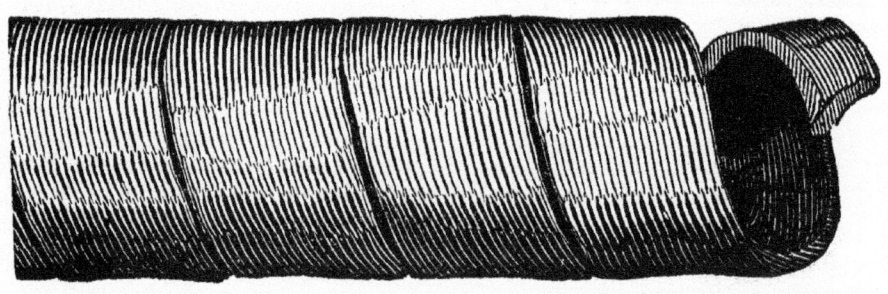

2.16 *A coil ready for barrel welding, from W. Greener,* The Science of Gunnery as applied to the use and construction of firearms, *Longman, 1841.*

THE MANUFACTURING OF THE BROWN BESS MILITARY FLINTLOCK

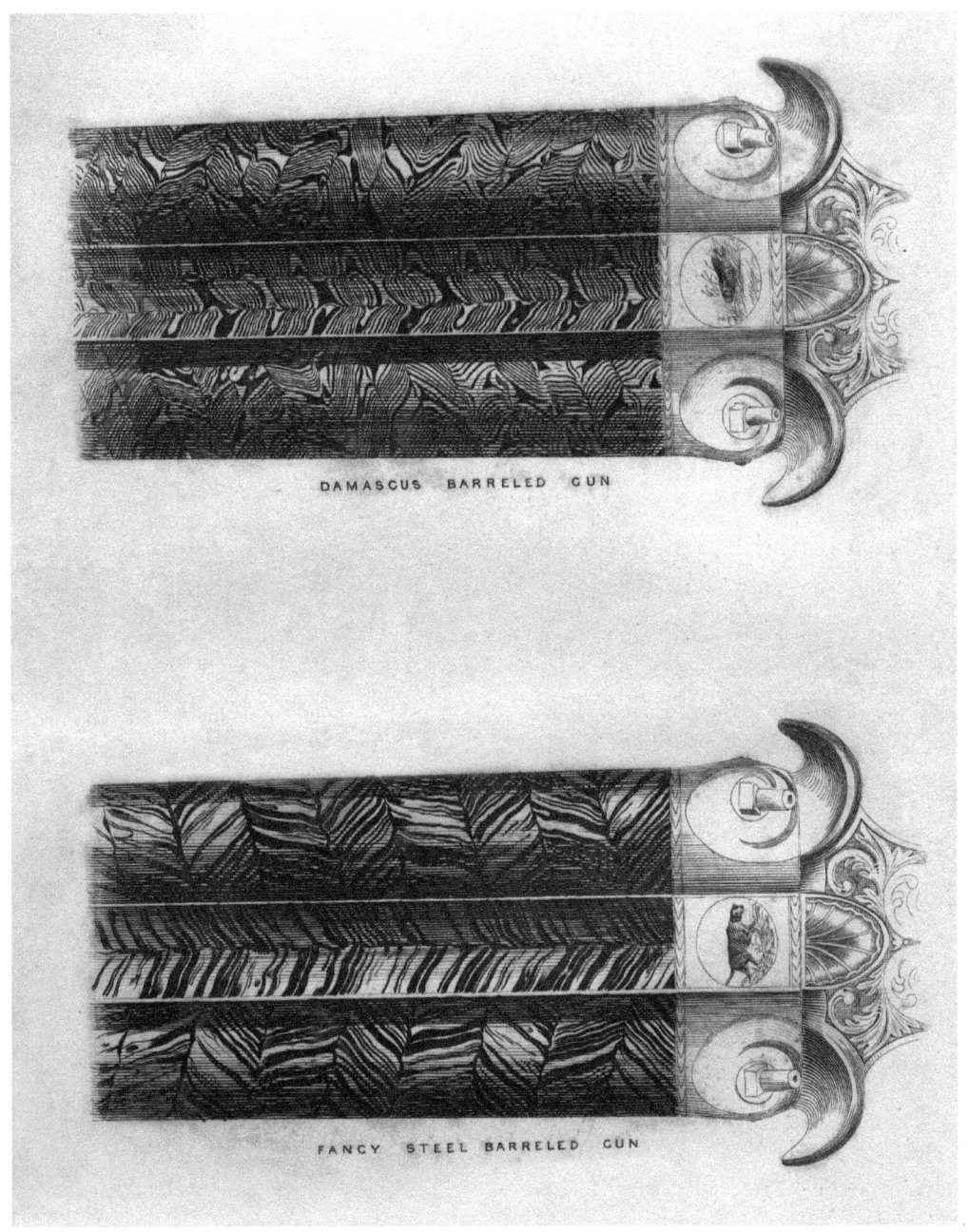

2.17 *Finished twist barrels from W. Greener, The Science of Gunnery as applied to the use and construction of firearms, Longman, 1841.*

THE MANUFACTURING OF THE BROWN BESS MILITARY FLINTLOCK

2.20 *Barrel boring in the Birmingham trade in the 1950s. (By permission of the Birmingham Museums and Art Gallery)*

Opposite Page

2.18 (Above) *Damascus barrel welding at W.W. Greener. (By permission of Graham Greener)*

2.19 (Below) *Barrel grinding, 1851. (By permission of the Birmingham Museums and Art Gallery)*

THE BIRMINGHAM GUN TRADE

2.21 *A close view of the barrel boring machine. (By permission of the Birmingham Museums and Art Gallery)*

Dry grinding was usually used, with serious side effects on health. Lathes were also used to turn the outside of barrels.[45]

The 1950 photographs in **2.20** and **2.21** show us an only slightly more modern approach to barrel fine boring or lapping in the post-war Birmingham trade. From the photographs we can see the use of line shafting-driven machines with weighted chains to provide the cutting force and feed, and fine and choke boring bits with steel cutters and wooden bore followers (**2.19**) that are packed out with paper to reach finished size. Note that single and double barrels are being lapped on the same machine. The machine bed appears to be wooden with metal facings on the guide ways.

2.22 is a photograph of the Grindery of William Hunt and Sons of Oldbury, also known as The Brades Co. Brades was a material supplier to the military trade. The passage of almost fifty years since the publication of the print of the grindery seems to have had little effect on the technology applied. **2.23** and **2.24** show their tilt hammer and gravity or 'drop' stamp shop respectively, both shops being driven by line shafting. Notice the jockey pulleys used to tension the tilt hammer belt drives and the number of two-man teams in the untidy stamping shop. These photographs are taken from their 1898 catalogue which says:

THE MANUFACTURING OF THE BROWN BESS MILITARY FLINTLOCK

2.22 *The grindery of the Brades Co., William Hunt & Sons Ltd, 1898.*

2.23 *The forge of the Brades Co., William Hunt & Sons Ltd, showing tilt hammers, 1898.*

2.24 *The stamping shop of the Brades Co., William Hunt & Sons Ltd, showing drop hammers in 1898.*

> Originally what small amount of mechanical power existed was derived from small water wheels, but these gave place to one of the earliest steam engines made by Matthew Boulton and James Watt, which in its turn, after about a hundred years of useful work, succumbed only a year or two ago… When the old wooden ramrods for military muskets were replaced by metal ones the steel for them was obtained from Brades and the name Ramrod Hall, which is still retained, was attached to the place they are made.

Brades also produced[46] large quantities of bayonets 'during the war with the American colonies'.

Wilkinson, also writing in his 1841 book, confirms the sophistication of the processes that could be used for musket barrel manufacturing. He describes the rolling in shaped rollers of 'skelps' or flat barrel blanks; and barrel welding by the hot rolling of cylindrical tubes with overlapping edges on a 'triblet' (cylinder), these welded tubes being subsequently further rolled between grooved rollers to their full length. The latter may have been invented by Henry Osborne,[47] a sword and barrel maker in Bordesley. It was supposed to be have been invented following a strike of barrel makers. The dates of Osborne's welding patents are noted to be between 1812/13 and 1817.[48]

THE MANUFACTURING OF THE BROWN BESS MILITARY FLINTLOCK

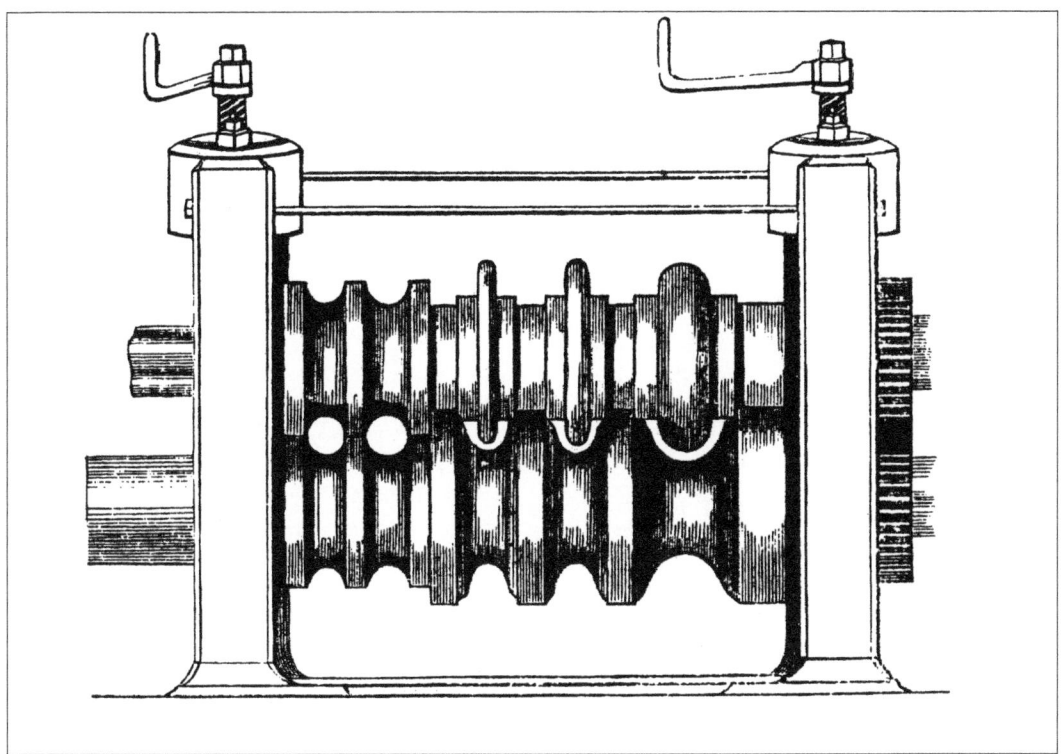

2.25 (Above) *A gun barrel bending mill stand, from Jervis White-Jervis, The rifle musket, A Practical Treatise on The Enfield Pritchett Rifle, 1854. (By permission of W.S. Curtis, Publishers)*

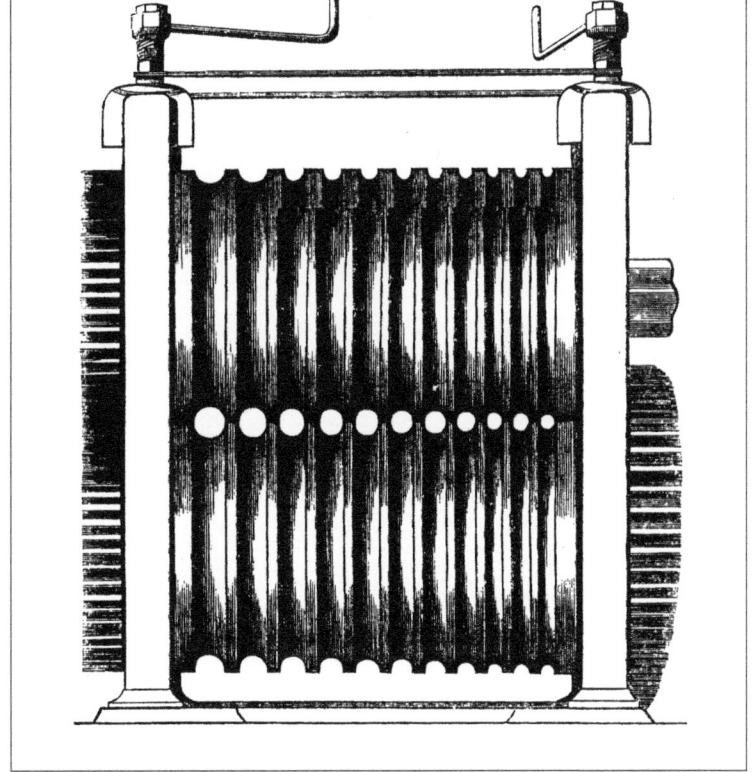

2.26 (Right) *A gun barrel rolling and welding mill stand, from Jervis White-Jervis, The rifle musket, A Practical Treatise on The Enfield Pritchett Rifle, 1854. (By permission of W.S. Curtis, Publishers)*

THE BIRMINGHAM GUN TRADE

There are also a number of earlier Birmingham patents on the use of grooved rollers in gun barrel making: John Jones in 1806[49] and 1809, Benjamin Cook in 1808 and John Bradley in 1811.[50] **2.25** and **2.26** show the barrel bending and rolling and welding mill stands subsequently used at Enfield to make Pattern 1853 barrels. Goodman, writing in 1866, also describes the process:

> Military barrels are made, in the first stage, by the process of rolling. A slab of iron 12 x 5½in, and ½in thick is first turned in a pair of grooved rolls until the edges meet. It is then brought to a welding heat and closed in a third groove of the roll. It is subsequently heated again, and passed thorough a succession of grooves on a mandril, until the 12in tube is drawn out to the required length of about 40in.

Goodman indicates that only two barrels in a thousand fail proof but more are rejected for flaws.[51] **2.27** shows the gun barrel proof process: barrels are loaded with an excess load, discharged and then inspected to see if they have failed.

Birmingham, the Centre of the Trade

At this time, Birmingham was the centre of the UK gun-making industry and would become the pre-eminent supplier of gun and gun parts to the remainder of the English

2.27 *Gun barrel proving, from W. Greener, The Science of Gunnery as applied to the use and construction of firearms, Longman, 1841.*

THE MANUFACTURING OF THE BROWN BESS MILITARY FLINTLOCK

2.28 *The Birmingham Gun Barrel Proof House.*

makers. For example, between 1804 and 1815, 1,743,824 military weapons, 3,037,644 military barrels and 2,879,203 gun locks were completed by the Birmingham gun makers and their Black Country suppliers, together with an estimated further million guns for the East India Co. and 500,000 sporting guns.[52] During this period the number of arms supplied to the government by the Birmingham trade was twice that supplied by London and exceeded the total output of France's ten government arsenals by more than 600,000.[53] The Galton family of Quakers were one of the most significant gun merchants, being able to make one gun a minute in 1772 with a 'prodigous amount for exportation'. Samuel Galton held Lunar Society membership and the family were well connected with Boulton, who supported them in negotiations with the East India Co.[54] The slave trade and later African trade took many – between 100,000 and 150,000 per year – of the lower quality Birmingham guns,[55] as slaving was going on from the 1690s to 1807.[56] The Birmingham Proof House was built in 1813, enabling local barrel proof to national standards essentially to filter out the lower quality usually unproofed Africa trade gun barrels. **2.28** and **2.29** show the Proof House today. The African Trade Guns have been described[57] as follows: 'The "male" gun was a superannuated "Tower" musket, or other comparatively sound weapon, tricked out with a coat of gaudy red paint to catch the eye' and: 'the "female" being a longer and lighter gun of indescribably vile quality, which was made new for the African market, and was sold to the trader for about seven shillings and sixpence. This last class of gun was essentially a Birmingham product'.

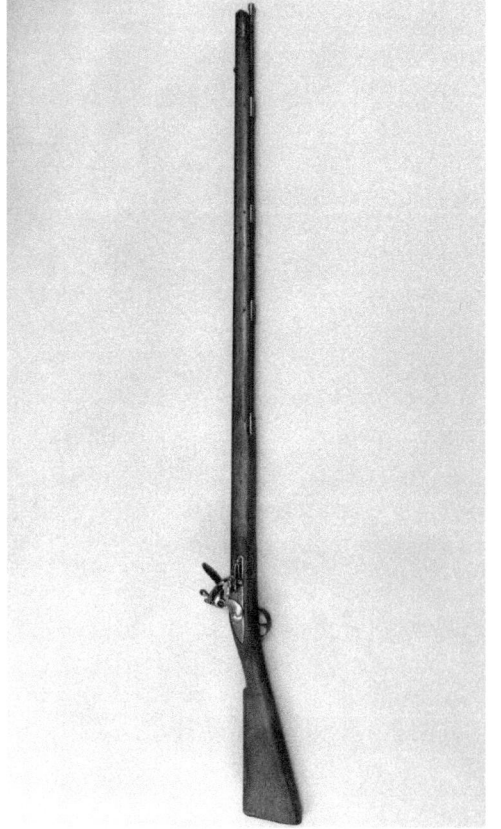

2.29 (Above) *The coat of arms of the Proof House.*

2.30 (Left) *A 'female' Africa trade gun.*

2.30 shows a female gun made by the last maker of trade guns, E. Barnett & Sons, the last of the London Minories gunsmiths, acquired from the Birmingham trade. The trade ended with the First World War.

It will be clear that the Birmingham trade had considerable success during the Napoleonic Wars with a significant fall off of orders after 1817. It has been observed that,[58] in such 'hostile' circumstances the normal rules of economics do not apply. One of the early classic textbooks, *The Economy of Manufactures*, by Charles Babbage[59] written in 1832, shows us the changing price of 'single roller', sporting gun locks as follows, the variation in price indicating the excess capacity in the industry after the end of the Napoleonic Wars.

Date	Price
1812	7s 2½d
1818	6s 0d
1824	5s 2d
1828	1s 10d
1830	1s 6d
1832	1s 11d

The Birmingham gun trade responded to a significant challenge[60] to supply the number of weapons required for the Napoleonic Wars; it made a fivefold increase in production from an estimated capacity of less than 3,000 guns per week in 1790 to 14,000 per week in the years after 1803. It did this without machinery – except for that used for gun barrel making – and without the innovations in other powered machinery demonstrated in the USA. This emphasises the elasticity of small producers with little capitalisation and the industrial structure required to meet irregular orders for military weapons from the British government. The trade responded to this extra demand by rethinking its 'custom' so that mass production could be achieved without the growth of the factory or sweated labour. As has been stated: 'the organisation of the Birmingham gun trade involved the interplay of three groups: the contractors, the small masters and the artisan-workmen who co-ordinated teams of apprentices within the workplace',[61] the most significant feature of production being the 'network of continuous negotiation between the three groups'. When an order was taken by the contractor he would negotiate with a merchant or the representative of the Board of Ordnance. The contractor would then negotiate with the small master for components of the gun, and the small master would in turn negotiate with the artisans and they with their apprentices – essentially a chain of sub-contractors. As all this was based upon fair prices, the negotiations extended over several weeks. However, when a large order had been received, the system could be very volatile because the small masters would not commit themselves exclusively to one contractor. The contractor would only be paid on completion of the order.

The trade expected irregular orders and intense periods of activity, balanced by periods of leisure.

The demand from 1803–1815 required a different approach. The trade took four strategies to meet the large wartime demand. It increased investment in gun barrel making plants; it established a contractor's cartel; it renegotiated apprenticeship regulations, particularly at production bottlenecks and introduced the bounty system in the workshop. The contractor's cartel was formed as the 'Committee of the Manufacturers of Arms and Materials for Arms' and handled all negotiations, with the Board of Ordnance representing contractors that were formerly competitors. One of the major issues was the supply of gun locks – this led to the committee renegotiating the apprenticeship agreements of twenty-seven lock-making firms so that: 'A person on having learnt the trade of forging will be entitled to two guineas and the person instructing him to three guineas, by applying to the Committee of Manufacturers of Arms and Materials for Arms at the Stork Tavern, Birmingham'. This relaxation allowed more to be trained while recompensing the skilled man for passing on his trade. The trained men saw themselves as independent workers in control of their own pattern of work and leisure – the bounty system was introduced in order to link them more closely to the masters. Bounty was a premium payment, of a maximum of £50 per year, taken in return for a skilled man binding himself to a small master for a period and delivering the required volumes of guns, perhaps in turn by taking on more apprentices. In this way the trade was able to meet high levels of production without high risk outlay on capital equipment. Later analysis has shown that the growth of many of the industries encouraged during the Napoleonic Wars was not subsequently sustained – this supports the conservative approach, taken by the Birmingham entrepreneurs, to capital investment.

Apprenticeship in Birmingham seems to have been a good deal more flexible than elsewhere[62] in the length of terms, types of training and opportunities available. Formal apprenticeships might only be a year (but could be much longer).

Other times of prosperity for the trade were the Crimean War (1853–1856), the American Civil War (1861–1865) and the Franco-German War (Franco-Prussian War) (1870–1871). The 1851 Census shows 5,167 gun makers in Birmingham and a private census from 1855 shows 6,840 employed, with a further 500 bayonet makers.

As with much local manufacturing industry in the eighteenth and nineteenth centuries, the Birmingham trade was the sum of activities of a large number of small masters, rather than running on the factory system – Boulton and Watt and their Soho Manufactury was 'a magnificent exception'. This contrasts with the large number of big factories that grew up in Manchester as cotton mills, for example.[63] The masters, while using machinery for barrel boring, did not use machine tools widely in the manufacture of locks and other parts, this being largely carried out by gun and lock filers. Complete guns were fitted rather than assembled.

Transport by canal to London and the new Royal Military Depot at Weedon Bec was important in the late 18th and 19th centuries and financing from gunmakers assisted the growth of the canal network.

The American Civil War

The American Civil War[64] lasted from 1861 to 1865. Slavery was at its core. The war was extremely costly in both lives – 359,528 Union and 258,000 Confederate soldiers died – and economically – it cost $15 billon directly, and $60 billon if losses due to physical destruction are added. It was perhaps the first modern, industrialised war. The Confederate States of America was created in Montgomery Alabama on 4 February 1861 following the election of Abraham Lincoln to the Presidency. The Southern States had said they would secede from the Union if Lincoln was elected. The war began on the 12 April 1861 when the Confederates fired on Fort Sumter in Charleston, South Carolina. The final surrender was on 26 April 1865.

Neither side was prepared for a war of this scale: the north and the south had a shortage of up-to-date small arms. High volumes of Pattern 1853 Enfields were made for both the north and south. The USA percussion muzzle loader design had been informed by the design of the Pattern 1853 and they took equivalent ammunition. Although their bores are apparently different at 0.577in and 0.58in, this is due to a subtly different method of measuring the bore of rifled weapons. Both first quality – interchangeable – Enfields were procured from the London Armoury Co. and second quality – non-interchangeable – weapons were procured from the Birmingham trade. Goodman[65] tells us that: 'The attack at Charleston on Fort Sumter commenced on 12 April 1861 and the first order for arms reached Birmingham the following 9 May.' His statistics show rises in proof and national exports. The numbers are the highest volumes achieved by the trade, which shows its elasticity. He estimates that 733,403 guns were sent to America from Birmingham and 344,802 from London – 1,078,205 in all over four years. Goodman again observes: 'The government can at all times rely on the resources of the arms trade of Birmingham in any emergency that may arise.'

The total number of weapons purchased by the north from 1 January 1861 to 30 June 1866 was 3,104,184 made up of 2,681,000 muzzle loaders and 423,000 breech loaders, including 89,653 Sharps and 106,667 Spencer shoulder arms. 801,997 of these came from Springfield, the rest from contractors.[66] James Burton, of whom we will hear more later, was in charge of Confederate Ordnance.[67] The well-reported Battle of Gettysburg was the turning point of the war:

> On that battle field it is said that 150,000 men were engaged and from the battle field after the engagement 27,000 muzzle loaders were collected. Of these it was reported that 24,000 were loaded and 12,000 double loaded. A large proportion were charged from three to ten loads.[68]

In Birmingham the income of skilled workers rose to fabulous figures: £20 a week was considered commonplace, in some instances barrel makers had earned as much as £50 a week. The Civil War 'led to much profligacy among the Birmingham gun trade workers. Stories are common of these men bombastically lighting expensive cigars with £5 notes in public houses'.[69] 'One of the local men, an expert percussioner, invariably came to work on horseback, stabling in horse in Whittal Street. Most of the workers, however, travelled to and from their work in hansom cabs; and its is related of one (for whose custom two cabbies had fallen out) that he engaged both by boarding the cab of the first, and tossing his hat into that of the second, instructed the driver to follow to the Gunmakers Arms; where great carousals nightly took place'.[70] At this time there were around 7,000 men working in the military gun trade alone.[71] We should also recall[72] that a considerable proportion of the output of the Birmingham civil trade went to America: 'and was there retailed by American gun makers whose names were frequently placed on the barrels or locks before leaving the Birmingham shops'.

In 1859 W.W. Greener[73] built his new factory at Rifle Hill, Aston. As soon as the business was established he began the manufacture of 1853 Enfields for both north and south – the whole work force was occupied night and day. Good prices were paid, up to 60s per rifle. Shipped via Liverpool, when the ships were ready for sale 'everyone would be employed throughout the night packing weapons and carrying the cases to the railway station'. The first order shipped to the USA seems to have had problems; when the order arrived at the Custom House in New York the value on arrival was much more than declared on shipping – this was eventually resolved but it was never determined who was generating the mark-up.

> The demand from the Northern States ceased in September 1863, supplies from their own factories being sufficient. Without notice the orders were withdrawn, as at the time the manufactories at Springfield and elsewhere were found sufficient to supply their wants. The smaller numbers subsequently made by the Birmingham trade must have found their way to the confederate states.[74]

Unfortunately, the Civil War also saw the introduction of the machine-made breech loader, requiring a precision – at a low price – beyond that of the Birmingham manual trade, the development of sufficient manufacturing capacity in the USA for it to supply its own needs and, at the end of the war, large quantities of military surplus.[75]

The Precision of Manufacture

Because of the nature of the manufacturing process used – fitting – it is difficult to determine the manufacturing tolerances used by the trade. The trade did not make to tolerance as it usually worked to a pattern. It essentially made individual components to fit individual weapons – a pattern weapon also did not allow exact comparisons of dimensions. However, a study of the range of sizes of locks produced allows us to determine an equivalent of tolerance or measure of the degree of interchangeability achieved by this manufacturing system. Indicative measurements of Ordnance lock lengths[76] allow lock *length* tolerance equivalents to be derived, for example +/-1/16in (+/-0.0625in) for the Pattern 1727 and + 1/16in −1/8in (+ 0.0625in − 0.125in) for the Pattern 1756 Land Service Muskets. The Pattern 1756 pistol lock that the average length is 6⅝in 'although rare aberrations as short as 6⅜in have been noted' – equivalent to a minus tolerance of ¼in (0.25in). It is worth noting that Maudslay with his Lord Chancellor bench *micrometer* could measure to 0.001in at the turn of the nineteenth century.[77]

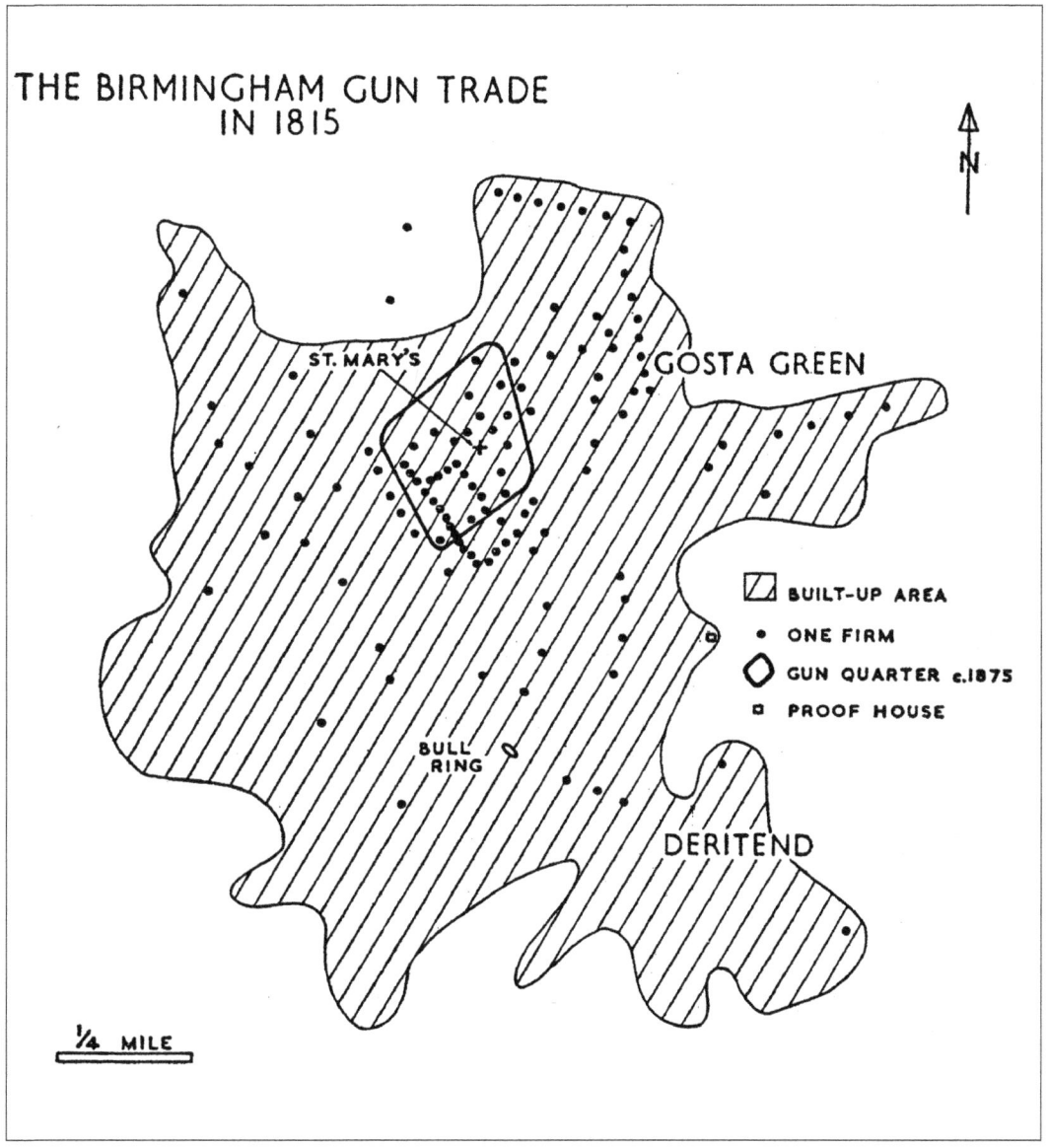

2.31 *The Birmingham gun trade in 1815 and 1875.*

THE BIRMINGHAM GUN TRADE

The Geography and Form of the Trade

Two excellent papers[78] allow the following description of the evolution of the Birmingham gun making trade (excluding barrel making) and its working practises. They particularly show its changing geography (**2.31**, **2.32** and **2.33**[79]), its concentration into the St Mary's area just to the north-east of the centre of the city (**2.31**), the transition to the more recent, post-1860, outer factory locations (**2.32**, – these are now in their turn disappearing) and the subsequent contraction of the domestic and factory-based trade located in St Mary's. By the 1740s a migration to the St Mary's area had begun. At this time, St Mary's was a select residential estate on the outskirts of the city. By 1829, three-fifths of the trade was centred on St Mary's church, thus creating a highly localised and specialised location which we would now know as a cluster. In 1925 only 19 of 203 firms in the gun trade were outside the St Mary's area, the 203 small firms concentrating on the sporting guns business.

Their original buildings began as the residences of the wealthier arms entrepreneurs of the eighteenth century. These men found it practical to use part of their homes as stores for components and finished weapons. They then became unattractive as homes and were turned over to the use of workmen, several working in

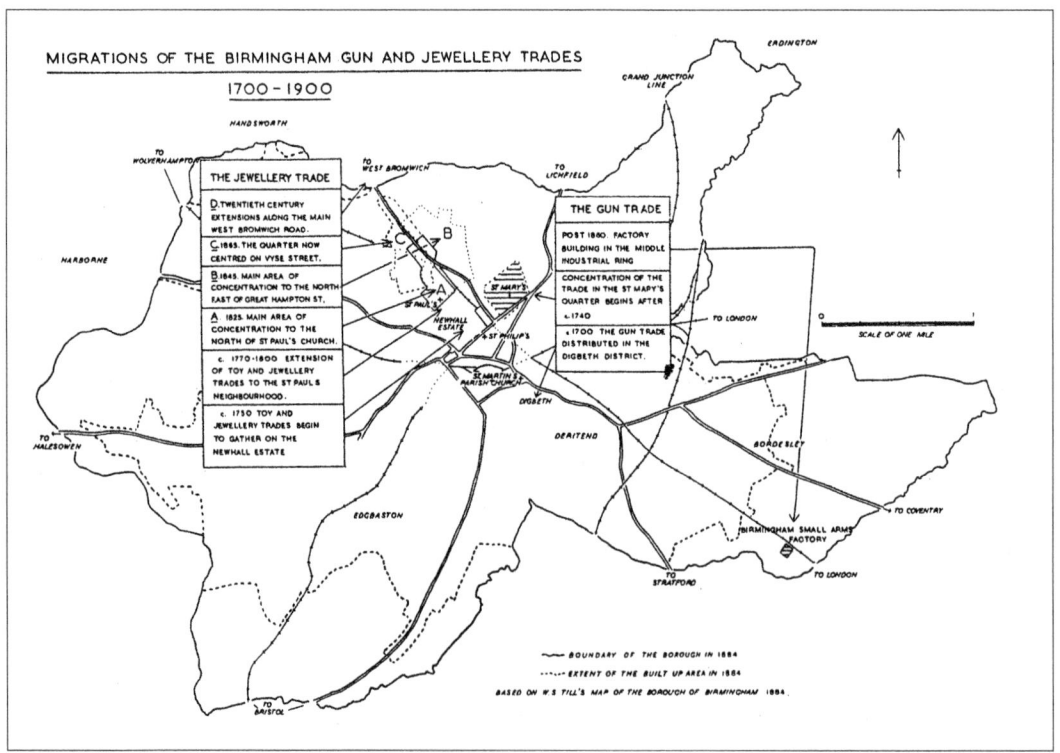

2.32 *The major movements in the Birmingham gun trade. (By permission from the Transactions of British Geographers)*

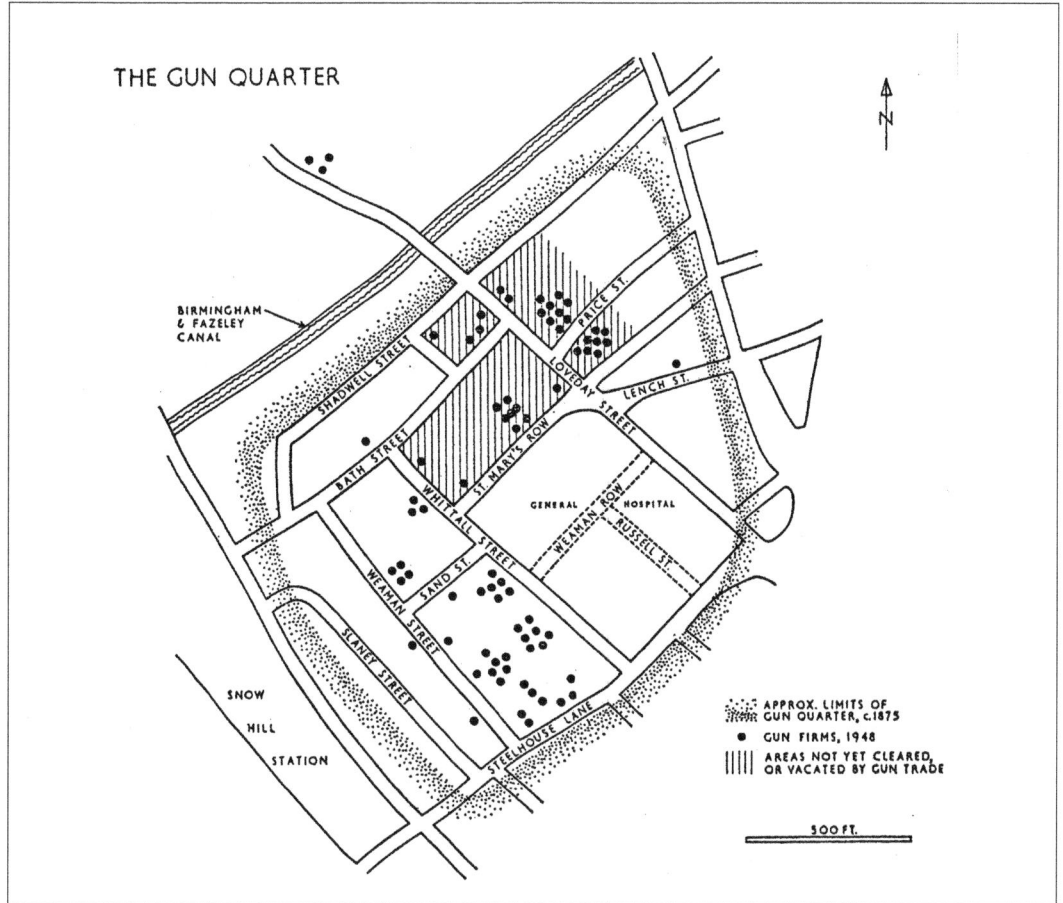

2.33 *The St Mary's trade in 1948 and 1964.*

one room. Skilled men, often specialists in a particular task, frequently hired from the entrepreneur or factor the use of bench space, a vice and lighting, known as a 'stand', and worked for the entrepreneur and others. The entrepreneurs frequently had further businesses, typically as publican, shopkeeper or even gentleman! A conglomeration of workshops tenanted by individual workmen therefore grew up, the workshops being frequently in the gardens of the original Georgian houses.[80] Both those who hired space and other outworkers[81] were sometimes called piecemen as they were paid by the piece. This was in contrast to true employees paid by the day who were called daymen. Entrepreneur gun makers – sometimes called factors or garret masters – could therefore have a mixture of daymen and piecemen, both true outworkers and those hiring space, working within their businesses. Young workers were used to transport components from one worker to another, adding further to its labour intensity. This pattern of work continues today in the sporting gun trade, though at a much reduced level.

2.34 *Tom Chambers of the Ideal Gunworks shaping a stock with a draw knife, 4 Price Street, 1964. Note the use of the wooden 'horse'. (By permission of Paul Chambers)*

THE MANUFACTURING OF THE BROWN BESS MILITARY FLINTLOCK

2.35 *Inletting the stock to receive the metal components.*

The Birmingham trade clearly employed a large number of people, mostly carrying out manual tasks in what is usually called the domestic system or sometimes the proto-industrial system (in contrast to the factory system). There were about 7,340 in the trade in the 1850s.[82] This, in combination with the numbers employed in other Birmingham metal trades, meant that there was significant labour elasticity – it was easy to move from trade to trade, although this could result in quality and supply problems – and that it required very little capital to set up in business, supporting the activity of the entrepreneur. In times of uncertainty these flexibilities were very valuable, the fortunes of the trade rising and falling with wars and other military and colonial activity.

The Bottlenecks

The major bottlenecks in the trade were the intensively manual and time-consuming stock making and lock inletting tasks[83] – see **2.34** and **2.35** for an insight into the manual processes of stock shaping with a draw knife and stock inletting – a key step in gun fitting – respectively. **2.34** shows the use of the 'horse', a wooden steady to stop the stock moving as it is worked on. Wilkinson however, observes in 1841:

> Machinery for the stocking of military guns has been proposed and adopted to a considerable extent in France and in America; the barrel, lock and furniture being entirely let into the wood by this means: but one inconvenience has arisen in the difficulty of filing the iron or brass work so accurately as to fit the wood thus prepared.

This is echoed in 1916 by Roe in his explanation of the evolution of machine tools. The solution of one problem also required the solution to another. The separation by the contractor system of the manufacturing processes of stock making and lock making also did not force either to consider their relationship with the other, thus aggravating issues with the evolution of the system.

A quote[84] allows us to see something of the people at the time. Matthew Boulton, towards the end of his life (he died in 1809), wrote:

> Birmingham was as remarkable for good filers and forgers as for their bad taste in all their works. Their diversions were bull baiting, cock fighting, boxing matches and abominable drunkenness with all its train. But now the scene is changed, the people are more polite and civilised, and the taste of their manufacturers much improved.

3

ARMORY PRACTICE OR THE AMERICAN SYSTEM OF MANUFACTURES

This chapter reviews the steps taken primarily outside the UK to develop the methods of interchangeable manufacture; in the United States the system that evolved became known as Armory Practice, in the UK we named it 'The American System of Manufactures'.

Progress in Europe

Some improvements towards interchangeability were made in the eighteenth century, particularly in France – interchangeable manufacture of guns was unsuccessfully attempted in 1717.[1] The French engineer M. Deschamps had claimed to be able to make flintlocks with interchangeable parts: 'exécutees avec tant de précision, qu'elles se convenoient toutes les unes aux autres'.[2] Réaumur, in his *Memoirs on iron and steel*, said in 1722:

> A musket with a broken barrel becomes useless because the lock, or the pieces of the lock cannot be fitted into another musket. But, when all the pieces are of the same diameter, the pieces of one musket can be substituted for another. A few broken pieces will no longer make all others useless. What is left of a badly shattered musket will serve to repair another.[3]

French armourers also said: 'In all the threaded holes, the same diameter which must be found in all of the muskets by virtue of which the screws from one lock plate will fit all of the other lockplates'.

Le Blanc of France, who famously showed his work to Thomas Jefferson in 1785[4] and Tipton victualler's son Henry Nock, when making his screwless lock, used templates or filing jigs to improve interchangeability. The screwless lock can be disassembled without using a turn screw or screwdriver. The Nock locks can be shown to be somewhat interchangeable but were not mass manufactured. Although hundreds or thousands were made, few appear to have been originally used: 'For some reason, whether the cost, or of the difficulty of making such a high grade lock in sufficient quantities'.[5] George Lovell, 'the foremost designer of percussion arms'[6] writes on Nock, after the event: 'These kind of locks were manufactured somewhat in the manner as the French "Platines Identiques" – the several parts were made by different artificers in Moulds, Gauges and Tools and fitted together by a separate set of men'. Le Blanc used, as part of his system the finishing of critical parts in dies reducing the amount of hand filing necessary, a technique he said had been developed in England.[7] He also used an extensive set of gauges.

The Swiss manufacturer Georg Bodmer used machinery to make gun locks in St Blaise in the Black Forest during the Napoleonic Wars, beginning in around 1806.[8] He stopped in 1816, perhaps because he was focusing on making the more complex flintlock rather than the percussion lock.[9] Bodmer is said[10] to have 'invented and successfully applied a series of special machines by which the various parts – more especially those of the lock – were shaped and prepared for immediate use, so as to insure perfect uniformity and to economise labour'.[11]

The Swede Christopher Polhem also made many innovations in mechanical engineering manufacture, including the use of simple gauges, as we will see later in the book.

Innovations in the United States

The next major steps were taken in the USA in the first forty years of the nineteenth century. These were in two areas: in woodworking machinery for stock making and in machines to allow interchangeable metal working.

These innovations were in spite of many actions[12] taken by the British government before the independence of the United States of America to constrain the development of manufacturing industry. These included the Acts of 1750 to ensure that: 'No mill, or other engine for slitting or rolling of iron, or any plating forge to work with a tilt hammer, or any furnace for making steel shall be erected'; and of 1785 to stop the emigration of mechanics and workmen in iron or steel and the export of any kind of machines or plans of machines. Clearly the success of the Revolution (1775-1785) and irritation at the earlier legislation spurred the Americans to innovate. Roe also tells us that a Scot, Hugh Orr, was the first to build a stand of arms in America, building 500 weapons in 1748.

3.1 *Blanchard's gun stock lathe. (By permission of the National Park Service, Springfield Armory NHS, Springfield, MA)*

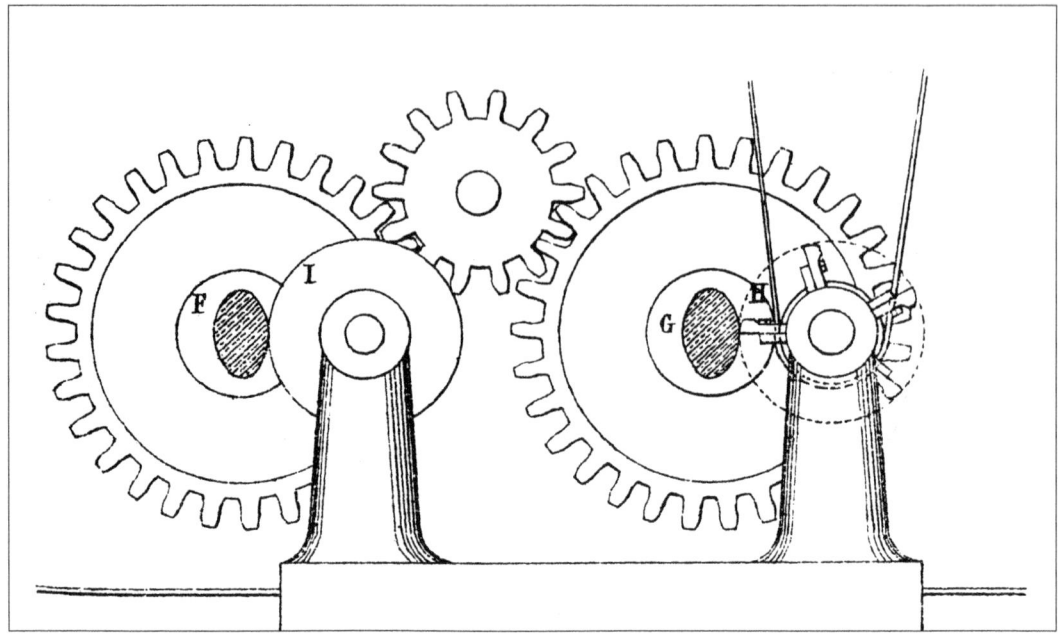

3.2 *Anderson's diagram of the principle of Blanchard's lathe, from the Proceedings for 1858. (By permission of the President and Council of the Institution of Mechanical Engineers)*

Working from the lead of the UK machine innovations, particularly the Portsmouth block making machinery, themselves inspired by Samuel Bentham working with Marc Brunel, in 1818 Blanchard developed the 'self directed' *copying* lathe for stock making, see **3.1** and **3.2**, clearly showing its principle of operation.[13] Completing a stock required seventeen machines. It was estimated that by using the Blanchard lathe, only seventeen men were required to produce 100 gun stocks a day in contrast to the seventy-five men needed for the traditional manual method.[14] The machines did waste a great deal of the raw material. This however, was not a problem given the good supply of wood in the United States, and perhaps one of the reasons for their neglect in England was because of the relative scarcity of wood.[15] It has been estimated[16] that savings due to the use of the Blanchard lathe was 25¢ per stock, not counting the capital costs of the machine, on a total musket price of $13.00 in 1820 – this is perhaps another reason why it was not accepted quickly in Great Britain. However; while the introduction of individual machines shows little savings, the introduction of the *System* kept prices stable by reducing the impact of labour costs.[17] It has been recently estimated that between 1808 and 1812 it cost $12.62 to produce a musket; by 1848–52 it was estimated to be $12.68 with a 91 per cent increase in wages, thus indicating a 27.5 per cent increase in productivity. **3.3** shows a contemporary 1863 photograph of a later variant of the Blanchard lathe, built by Ames to the design of Cyrus Buckland.[18] The evolution and novelty of Blanchard's lathe is described in detail by Caroline Cooper,[19] who identified that he made sig-

3.3 *A contemporary photograph of the Ames variant of Blanchard's lathe. (By permission of the British Library)*

nificant innovations in machine kinematics to allow the copy turning of objects that have very significant differences in their dimensions around the axis of rotation – as does a gun stock – and that he applied his machines as part of a system to increase overall efficiency.

There has been debate as to where interchangeability was demonstrated and by whom – increasing sophistication of mechanism function and changes in manufacturing processes and measuring methods have also continually redefined the level of interchangeability expected and achieved.[20] Eli Whitney has been given the honours in the past, however Rosenberg[21] gives a full and unbiased account. He states: 'The notion that the system of interchangeable parts sprang full-blown from Whitney's genius in musket manufacture has now been given a decent burial'.[22] Since the mid-1960s there has been a progressive re-evaluation of the roles of individuals in the development of the key technologies of interchangeable manufacture in North American armouries and an increased recognition of the relative contribution of techniques originating from Europe by the transfer of people from the United Kingdom.[23] The techniques with their origins in gun making that led to interchangeability – particularly jigs, fixtures, gauges and the milling machine – 'were developed and refined over many years by the crafts people in several New England armouries'. These innovations followed the placing of orders by the government at the turn of the nineteenth century that encouraged interchangeability because of its value on the battlefield.

Rosenberg writes:

> Throughout the whole of the first half of the nineteenth century and culminating perhaps with the completion of Colonel Colt's armoury in Hartford in 1855, the making of firearms occupied a position of decisive importance in the development of specialised precision machinery... it is clear that both Eli Whitney and Simeon North employed crude milling machines in their musket producing enterprises in the second decade of the nineteenth century as did John D. Hall in the Harpers Ferry Armoury.

Whitney recorded his use of milling in 1823 and Ethan Allen first used formed cutters for milling parts of pistol locks in 1830.[24] Deyrup, in her germinal book, *Arms Making in the Connecticut Valley*, written in 1948, notes that the use of power rather than hand-driven milling machines in the 1830s gave a significant impetus to being able to practically achieve interchangeability. More recent work[25] suggests that the milling processes used until the 1840s were closer to the hollow milling process used in the tumbler mills seen in Chapter Two because of the deep chatter marks found on the artefacts produced and shows that the machined components, particularly tumblers with their complex geometry, had a considerable amount of hand finishing to size even until the 1880s.

In 1828 North[26] extended the principle of interchangeability to guarantee that 'the component parts should be interchangeable, not only in the lot contracted for, but that they may be exchanged in a similar manner with the rifles made or making at the national armouries'. Tolerances of Hall's breech loader, adopted by the USA Army in 1817, were measured in thicknesses of paper – the gap between the barrel and the chamber could admit one sheet of paper but not two, this tolerance is estimated[27] to be approximately 0.006-0.008in. Hall's breech loader was made in both Harpers Ferry and North's Middletown armoury,[28] hence the comment that rifles could be interchangeable between armouries. Hall also recognised, as did Maudslay,[29] the importance of a measuring datum or 'bearing' on the component being manufactured.[30] Gordon[31] measures the average deviation from the mean of tumbler components in the early years of Amercian manufacturing as follows: Model 1795 (made in 1806), 0.012in; Model 1816 (made in 1835), 0.004in; Model 1841 and 1842 (made in 1850), 0.002in; and Model 1873 (made in 1880), 0.001in.

Returning to Rosenberg's commentary: 'The subsequent development of the "manufacturing" or plain milling machine was largely the development of the national armouries, especially the work of Thomas Warner at the Springfield Armoury, and such gun producing firms as Robbins & Lawrence'. Robins and Lawrence[32] was formed in 1844 when young gun maker Richard Lawrence joined forces with the more experienced Nicanor Kendall and Samuel Robbins, a retired timber merchant, to bid to supply 10,000 1841 Harpers Ferry rifles for the Mexican War. Their bid was successful and they were contracted to supply the weapons at $10.90 each within three years. They drew a work force of 150 skilled mechanics around them and the expert Richard Lawrence led the construction of the machinery. They fulfilled the original contract within eighteen months and were contracted for a further 15,000.[33] Kendall consequently withdrew. Robins and Lawrence subsequently manufactured and perfected designs of the Jennings and Sharps rifles, ultimately exhibiting their manufacturing machinery at the Great Exhibition of 1851, which will be discussed later.

3.4 shows a Robins and Lawrence-like manufacturing miller in the collection of the Windsor Precision Museum; this is housed in the original Robins and Lawrence building. **3.5** shows a form cutter for cutting lock plates that can be sharpened without losing its form. Subsequent UK visitors to the USA[34] emphasised the importance of managing milling cutters and gauges, as did Goodman in 1866: 'One of the chief points to be attended to is the keeping such tools permanently to the original form; for if there is any departure, however small, the several parts will not interchange'. The form of milling cutters was replicated using a standard cut edge to create duplicate milling tools. They were also sharpened when hard 'on a small lathe driven at high velocity, are placed small circular cakes, composed of glue and fine emery… these are of various sizes and forms to suit different cutters' – form grinding. Roe[35] relates the businesses of the early gun and machine makers to show their connections (**3.6**). Such interactions are shown by the purchase of a Robbins & Lawrence machinery plant in Hartford, originally constructed for them by The Sharps Rifle Co. when the Robbins & Lawrence business failed, due to speculation in railway

3.4 *A Robbins & Lawrence-like manufacturing milling machine in the American Precision Museum. (By permission of Peter Smithurst)*

machinery and its late delivery of weapons to Great Britain.[36] Christian Sharps himself worked for Hall for a number of years at Harpers Ferry.[37] Another author[38] has observed that there were few innovators and that they were all related by work or marriage! Many of these machine makers were significant exporters; Pratt & Whitney for example provided gun-making machines for armouries in Spandau, Erfurt and Danzig.

Harpers Ferry, John Hall and Interchangeability

Woodbury, writing in 1959,[39] particularly begins to highlight the contribution of Hall at Harpers Ferry as recorded by James Carrington and Luther Sage in 1827, after their visits to assess the armory. Later investigators[40] focussed on the physical evidence on the surface of the artefacts produced – the 'witnesses' of the manufacturing processes being used to resolve some of the ambiguities in the historical record. The key technologies of interest in this discussion are the milling processes – the production of flat surfaces using a multi-edge cutter – its precision and the machines used. In the late 1960s and 1970s a number of people,[41] building on earlier work,[42] began

3.5. *The formed milling cutter set up on the Robbins manufacturing milling machine. (By permission of Peter Smithurst)*

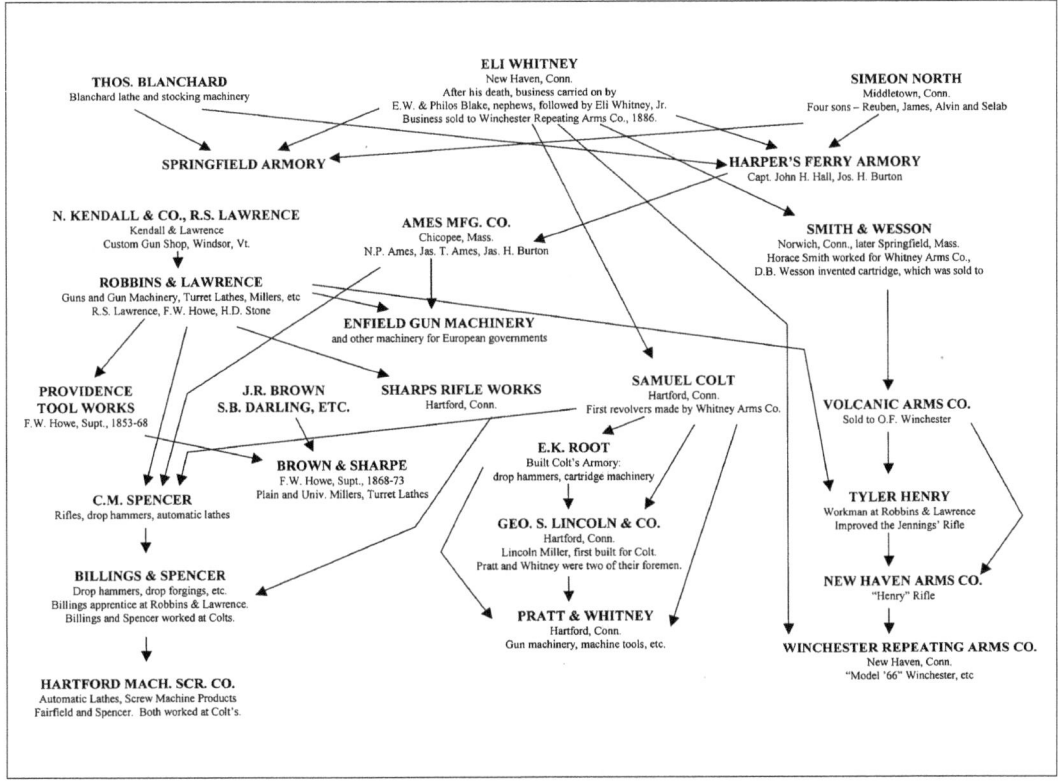

3.6 *The relationships between the early American gun and machine makers, adapted from Roe, 1916.*

to articulate that the contributions came from collaborations between individuals, for example Hall of Harpers Ferry, Nathan Starr and Simeon North of Middletown and individuals at the Springfield Armories under the encouragement and guidance of government procurement officials, strongly influenced by the French. The site at Springfield formed the springboard for many later innovations and their wider dissemination from arms making into other industries – and to Europe via machine makes such as Ames and Robbins & Lawrence. It has been particularly identified[43] that the economics of manufacture are modified in the case of military arms by other drivers – manufacturers for civil arms production may have to modify the techniques they use to make them affordable[44] and as may military manufacturers outside the national armoury system.[45] This identifies why the emphasis on the expensive techniques of interchangeable manufacturing may have started where they did. Recent accounts[46] recognise the particular contributions of Hall at Harpers Ferry to the early stages of this process.

The first superintendent of Harpers Ferry in 1798 was Joseph Perkin, who had served an apprenticeship in the Birmingham gun trade before emigrating to America in around 1774. Perkin was the creator of the Pennsylvania-influenced long-barrelled small bore Model 1803 military rifle. He was succeeded by Marine

T. Wickham, who subsequently went into business as a supplier of material to both Harpers Ferry and Springfield Armories. During this period the practices in the factory echoed those of the Birmingham and Pennsylvania gun trades – artisans could come and go as they pleased, used hand tools and were divided into six disciplines: barrel making, lock forging, lock filing, brazing (furniture making?), stocking and finishing.

In 1808 the armory was required to tool to produce 15,000 muskets annually. James Stubblefield, in charge at the time, visited the Springfield and Whitney armouries and brought back the additional division of labour and piecework (payment by the piece) applied at Springfield, putting it quickly to effect with fifty-five specialist tasks established – occupational specialisation.[47] The only task unaffected by division of labour was stock making. Specialisation of labour also accelerated at Springfield with 100 different occupations. This occurred with little innovations in the techniques used. Springfield, Whitney and Harpers Ferry used jigs and tumbler or 'hollow' mills in lock making. No gauges other than perhaps plugs were used until 1821. In retrospect there seem to be similar approaches taken to the processes in France, England and the United States.

Significant changes in the techniques used happened in 1816–1817. In 1811, John Hall had patented a novel design of breech loading rifle,[48] Hall considered that these should be made on the interchangeable principle 'using labour-saving machinery' and lobbied for their acceptance by government. Harpers Ferry began to make them in 1818 under Hall's supervision as Assistant Armourer from April 1819. Hall began to construct the Rifle Works and tools and machinery, completing this and the associated 1,000 rifles contract in December 1824. The machinery had been constructed to Hall's design, mostly to 'establish methods for fabricating arms exactly alike'. The government sought both to establish the value of Hall's breech loader and to assess the techniques and machinery that had been used to build it. James Carrington was critical to this: he was well known as an armourer and machinist, had served as a foreman at Whitney's manufactory from 1799–1825 and had acted as an independent inspector of arms. In December 1826, Carrington and colleagues spent three weeks examining Hall's machinery. They reported that the machinery was novel for both cutting iron and steel and woodwork, that there were no identifying marks on the parts and that the parts would interchange after hardening. Hall had given the first true demonstration of interchangeability; even at the Springfield armoury components had to be marked. Subsequently, such interchangeable rifles were also made by North at Middletown in 1834 which were also interchangeable with those at Harpers Ferry but this required much work on gauges.

Roswell Lee, superintendent at Springfield between 1815 and 1833, had begun to experiment with gauges in 1817 and had reached a point of sophistication using hardened steel gauges 'for the particular part' by 1819. Such gauges were subsequently introduced at Harpers Ferry in 1823. Lee was driven to exchange his practices by personal zeal and by the efforts of Lt-Col. George Bomford, who had the responsibility to promote efficiencies at the national armouries. Hall particularly used three sets of sixty-three gauges for his breech loader – one for his workmen, one for inspectors and one master set that he held in his office as a control that was not subject to hard usage.

Hall's breech mechanism had twenty-three parts with complex geometry and required a number of deep and irregular cuts. Hall created machines working from the innovations of others – including steps in the creation of the powered plain miller – but seems to have been more attentive to balancing the drives of his machines and constructing them to massive proportions to contain the vibration generated by the cutting process. Hall seems to have taken the significant step of engineering the product and the manufacturing process together; he particularly uses prismatic – 'rectangular' – product features capable of being made by the milling process that are readily measurable.

In hollow milling, cutting takes place on the circular face of the multi-edge tool, the hole through the tool allowing the passage of a central pin left on the work piece. A special purpose hand-driven hollow milling machine was frequently used in Europe for machining tumblers (recall **2.8** and **2.9** from the nineteenth-century Birmingham trade). An edited translation[49] of a description by Gamel, made in 1827, shows the many original techniques brought to the Russian Royal Armoury at Tula by the Birmingham man John Jones in 1816.[50] Gamel shows both a hand-powered tumbler mill and a small belt-driven tumbler mill. Less sophisticated tumbler mills – not using circular cutters – are shown in Diderot's *Encyclopedia* of 1751–65.[51] Hall's key step appears to have been to use well-designed powered milling machines capable of executing a number of different milling processes.

It has been identified[52] from the Carrington report that Hall had three forms of milling machines, two of them novel. James Carrington is usually regarded as a neutral and expert witness because of his earlier employment by Whitney. Carrington describes the machines in his report of 1827. The machines were described to be belt-driven from balanced pulleys (minimising vibration and chatter) and of massive style or construction using metal castings,[53] (again minimising vibration and chatter). Chatter is the surface defect produced when the cutting tool vibrates. The three machines are described as a 'straight cutting, curved cutting or lever cutting'. The lever cutting machine has been described as 'similar in form to some hand milling machines now in use'.[54] The straight cutting machine is usually considered to be a plain (or manufacturing) miller, similar to that used through the twentieth century[55] and the 'Whitney Miller', discussed later in the chapter is now considered likely to derive from the machines of Hall. The curve-cutting machine is a bridge – essentially a copy mill or tracer –described by James H. Burton:

> Bridge milling machines employed a fixed table which held the workpiece in position in its upper side and a copy of the required pattern underneath. The pattern rested on a fixed fulcrum or bridge situated below the table and in line

with the revolving mill cutter'... when set in motion... the bridge traced the pattern [to copy it].[56]

Carrington also noted that Hall kept a log of the cutting time for components to drive improvements. Importantly, these techniques will have travelled with contracts to manufacture weapons interchangeable with these at other armouries.

Such techniques allowed Hall to dispense with filing jigs and use boys to attend the machines. Hall's influence was also extended by the number of men who worked with him or that he trained, who then moved to work in other arms plants. These included many who went to the Springfield Armory. Hall's influence in the creation of plain milling also seems to have propagated from Harpers Ferry to North at Middletown and from there to Springfield; from Springfield it went to Ames and Robbins & Lawrence and subsequently to Great Britain.

Making the Hall breech mechanism was perhaps the most complex precision manufacturing task attempted in the United States before the mid-nineteenth century[57] and it has been concluded that the block component of the Hall breech was milled and hand finished.[58]

Gauges

Although gauges were used in manufacturing in the UK (lock receiving gauges used by the East India Co. in 1798[59] and 1818[60] to check the fit of assembled locks to stocks still exist), one of the major differentiators between USA and UK practice was the use of large suites of gauges, rather than the use of sealed patterns. Gauges were thought absolutely necessary because it was: 'Only by means of a continual and careful application of these instruments that uniformity of work to secure interchanges can be obtained'. This is taken from the commentary on the tenders arising from the visits to the USA in 1854, described later. Springfield used hardened gauges in 1819.[61] Receiving gauges are said to have been used at North's Middletown factory in 1829, and were regularly in use at Springfield by 1840.[62] The total number of gauges in the Springfield inventory are as follows:[63]

Year	1834	1835	1838	1839	1842	1844
Number	382	482	566	666	754	754

During this period the armory headcount remained at around 250 and essentially only one model of arm was made; by 1848 practical interchangeability was achieved. Significant steps were made at Springfield towards this new level in the 1840s. These included the acquisition of a set of weights and measures standardised by Ferdinand Hassler, head of the Bureau of Weights and Measures of the Treasury Department, to aid the making of model percussion arms. It has been emphasised by Deyrup, writing in 1948 before some of the new understanding that we have on the contribution of Hall,[64] that Thomas Warner, the Master Armourer of Springfield, obtained uniformity at a new level – removing the need to assembly locks in the soft state and file them before hardening into marked assemblies or to sets of assemblies that were interchangeable within batches of ten – by the use of milling machines, by jig filing and by careful inspection between processes.

Significantly, the gauges for the 1841 rifle were made largely by hand filing, except for holes and cylindrical surfaces.[65] Receiving gauges[66] were later commonly used in the interchangeable manufacture of small parts. They check the shape of a usually complex-shaped component by seeing whether they fit within a boundary. Many of Maudslay's pupils were active in their use of gauges during this period, perhaps learning from his practices, for example Richard Roberts in 1825 and Whitworth. Whitworth gave his key paper on screw thread standardisation and measurement 1841 at the Institution of Civil Engineers and, in 1846 laid down that his staff work to three dimensional (first angle projection) line drawings[67] and that they checked each piece with a difference gauging system. Roe[68] makes the key observation that the general use of accurate working gauges, particularly in large sets, which were hardly known in England, developed rapidly in America. As we have indicated, the receiving gauge checks the external form of the component to ensure that it will fit with another component. However; it cannot determine the characteristics of the function of the two components unless it is constructed as a limit gauge.[69]

Why did it Happen in the United States?

Because of the limited number of skilled hand craftsmen in North America, the interchangeable American System of Manufactures was developed between 1800 and 1840 in the armouries of New England, hence its USA label of Armory Practice. The key elements of this system for the manufacture of military firearms were the use of special purpose machines in place of skilled craftsmen and the evolution of the new disciplines required to make small precise interchangeable components. Specialisation or division of labour was used in both the UK and American industry; in the UK a single workman was specialised to a single component carrying out all the operations on that part, whereas in the USA a workman performed only one or two operations on the part.

Perhaps the major reason for the USA to go in this direction of creating machines instead of developing skills, was the lack of large numbers of skilled people: there was no labour elasticity. During this period, the UK was focussed technically on heavy capital goods: large prime movers and locomotives and consequently large machine tools. It is also worth noting that the Institution of Mechanical Engineers was founded in Birmingham in 1847[70] and was largely focussed on such large machine issues. As understanding capitalisation is of major importance in this discussion, it has been noted[71] that the primitive accounting methods of the early American businesses, for example that of Simeon North, could not accommodate or recognise the need to depreciate the new capital plant. The little profit really being made was not therefore visible, nor was there drive for reinvestment in plant. Others comment that the market place in the UK was not conducive to interchangeable manufacture – Britain's gun makers did not have the turbulent frontier of the United States nor the encouragement of government.[72]

How did the Americans Organise Manufacture?

In closing this section it is of value to review some North American organisational practise – internal contracting – in this era which has some similarities to the autonomy expected among the members of the Birmingham trade. The system has been described[73] in operation in the 1870s at Winchester. With the growth of larger manufacturing plants with the coming of machinery there was an increasing requirement for specialist management because the capitalist providing the funds rarely had the requisite technical, innovation and process improvement skills to lead production. This led to a system of expert internal contractors: super-foremen. These men hired and fired their own workers, trained their own apprentices, managed the job and supplied components to the company. The company supplied raw materials, the use of floor space, machinery, light and power. The management credited the contractor for the finished work which passed inspection, and debited him for the wage costs of his men. The contractor was also paid a wage. His reward was a combination of a wage and profit (or loss) from his department's production. The company used its own employees for maintenance, inspection – i.e. quality control – acceptance or viewing as it was known in the UK, and assembly. The system is said to have begun in Nathan Starr's Middletown plant in 1798 and was employed by Eli Whitney, Robbins & Lawrence, Brown & Sharpe, Colt & Remington and within Harpers Ferry and Springfield Armouries. Blanchard was employed as an inside contractor at Springfield between 1823 and 1827.[74] Inside contracting played a significant role in the development of the master tool builders because the development of more efficient machinery by the contractor increased their profit.[75] The system declined in the early years of the twentieth century. This was as a result of less innovation being required from the contractor with the stabilisation of the product and manufacturing technology, a drive to reduce costs, the introduction of Taylorism[76] – 'scientific management' – and tensions between management and contractor because of the economic and social positions of some of the more successful contractors.

Whitney's First Milling Machine?

According to the distinguished machine tool historian Woodbury:[77] 'Whitney's first milling machine was rediscovered in 1912 through the efforts of Joseph Roe – it was identified by Eli Whitney's grandson of the same name as having been pointed out to him since boyhood as built by Eli Whitney and the first miller ever built.'

Roe seems to have dated it at 1818, simply from the statement in the *Encyclopaedia Britannica* that: 'The first very crude milling machine was made in 1818 in a gun factory in Connecticut' and assuming that this could refer to no other than the Whitneyville plant. This gives us an understanding of the need to apply a robust methodology to working out the genus of the milling machine! Woodbury states: 'From the evidence and its design features it is the oldest milling machine extant, as well as the most significant in its influence on later machines'. Edwin Battison has published widely[78] on the machine to try and establish its real genus to suggest that it actually is from Hall's rifle works at Harpers Ferry and should be dated at 1827. Woodbury repeats Roe's original 1912 description of this legendary machine as we do here:

> It stands about 18in high and is about 2ft 6in sq overall. The base is a solid wooden block carried on short wrought iron legs. The main spindle, about 2½in in diameter, was driven by a single pulley, which must have been about 20in in diameter, as the wooden base was gouged out to clear it. The spindle ran in solid soft-metal bearings in two flat housings, bolted to the square box like frame. The connection between these uprights and the frame is so rough that it does not seem, even allowing for ninety years of rust, to have ever been a machine fit. On the inner end of the rear bearing is a wrought iron plate, which engages in a groove on the spindle which controlled its position lengthwise.
>
> …The slide,[79] jam and gib-bolts are lost. The slide ran between a gibbed bracket, cast on the frame, and one of the V's in the front bearing support.
>
> The feed was taken from the double-grooved wooden pulley between the spindle bearings down to the lower shaft by a round belt, the return side of which ran up through a hole in the wooden base… a wooden cam, two wrought iron plates and a swinging bearing carried the rear end of the worm shaft, the front end of which could be raised and lowered.
>
> The short bearing between the handle and the worm was carried on a small vertical slide, which had a spring latch to hold it in the engaged position. By dropping the latch, the worm on the feed shaft was disengaged to allow hand feeding.
>
> The feed screw was fairly well cut with what was originally a square thread, but along the middle, under the cutter, it has been worn to a sharp V. The keyways are rough chipped, and some, at least, of the bolts appear to be hand-made.
>
> A glance at the worm wheel… shows that it was made before the days of involute teeth. It is made of wrought iron and the worm is brass… It is a

quaint little mechanism, but the smile it provokes usually dies away in very genuine respect.

The Use of the File in the Armory Practice

Recent authors, particularly Robert Gordon[80] and Patrick Malone,[81] have examined the technologies of Armory Practice by looking at the metal artefacts produced with a more disinterested eye than was usual by earlier authors. Gordon shows that Hall finished milled surfaces with the file to achieve gauge dimensions. Machines were used for roughing components from forgings and files and filing jigs were used in the hand finishing process. The performance of Hall's machinery, in spite of it being physically robust, was limited by the cutters available. Gordon has also shown,[82] by the examination of a significant number of tumblers from well-preserved arms, that components of the M1842 Springfield Musket – with demonstrably interchangeable parts – were finished to gauge with files. Also the gauges used for the M1841 Rifle were soft (except for barrel plug gauges) and brought to final dimension by filing.[83] The practice of hand filing tumblers was still in place in 1884. When Springfield was newly commanded by Maj. Alfred Mordecai in 1892, one of his major tasks was to move away from inefficient practices, including the filing of parts to gauge by hand.[84]

Many of the steps made later were incremental and involved the use of jigs and fixtures – called 'little kinks and devices' by Fred Colvin, a prominent early twentieth-century American Machinist engineer journalist – for removing filing stages for soft components. Even the Krag magazine rifle, adopted in 1892, had considerable hand finishing and fitting, and final fitting with a whetstone was required until the production of the M1 rifle in the 1940s. Many of the gauges used in later rifles were developed from earlier generations.

As we have already indicated, accounts, including more recent ones,[85] of the manufacture of small arms, indicate that commercial concerns, such as Winchester and Remington did not seek to achieve the level of interchangeability of the government armories. It proved more economic because of higher productivity to hand finish weapons than try to make all components completely interchangeable.

The American System of Manufactures

Armory Practice, The American System of Manufactures has a number of features: the use of powered machines in place of hand power, the transfer of skill from man to machine, the use of piece work-based reward systems, expert internal contractors and the most novel part of the system,[86] the use of techniques to make interchangeable parts. It arose in military gun making in the United States but built upon many European innovations. One of the most significant steps was taken by an individual, Blanchard, but many of the steps, particularly in metal working, were incremental steps and were the work of many people – some of whom, with their supporters, were better at promoting their role than others. In this book we are focussing on how the techniques of interchangeable manufacture made an impact on the traditional international centre of the handicraft-based gun-making industry, Birmingham.

4

ENGLAND EMBRACES THE AMERICAN SYSTEM OF MANUFACTURES

The changes in the United States were very apparent in the UK, following the Colt and Robbins & Lawrence exhibits at the Great Exhibition. This led to a government review and a number of visits from England to arms manufacturers in the USA. Following this, a new generation of manufacturing equipment was installed in the Royal Manufactory at Enfield as an example to the British gun trade.

Colonel Colt Comes to London

Samuel Colt had a major influence on the UK's approach to gun making. Colt, an ambitious inventor, businessman and self publicist, included a startling display of guns in the Great Exhibition of 1851 and gave a lecture to the Institution of Civil Engineers in 1851.[1] Charles Manby, Secretary of the Institution at the time, later had commercial interests in Colt's enterprises and became his friend and agent – Colt had targeted him as a mechanism of getting attention.[2] The lecture was entitled: 'On the application of machinery to the manufacture of Rotating Chambered-Breech Firearms and the peculiarities of those arms'. The lecture gave rise to such interest that the discussion carried over two evenings. While the lecture is well-known, the manufacturing aspects are usually under emphasised.

In the lecture, Colt gave some indication of the manufacturing approach used at his Hartford factory but was cryptic over the detail. He identifies that machinery is used for: 'eight-tenths of the whole cost of construction [being] induced gradually to use machinery to so great an extent, by finding that with hand labour it was not possible to obtain that amount of uniformity, or accuracy of the several parts... obtaining uniformity as well as cheapness'. The direct cost of one of Colt's revolvers was $6.³ He also notes the ease of replacing parts and: 'on active service a number of complete arms may be readily made up from portions of broken ones, picked up after an action'. The guns are made by:

> Hundreds of distinct operations, involving a great variety of peculiar contrivances and mechanical motions [because] in America, where manual labour is both scarce and expensive, it was imperative to devise means for producing these arms with the greatest rapidity and economy, and at the same time with such uniform precision, as could only result from the use of self acting tools...The machinery... though apparently intricate... is in reality composed of the simplest elements and consists in a repetition of known mechanical actions specially applied.

Starting with a description of the manufacture of the lock frame, Colt takes us through some of the key processes used: 'Like all the other parts, the lock frame is [hot] forged by swages, and its shape completed by one blow'.⁴ We should note that the lock frame, while smaller than in some designs of revolver, is a relatively large forging when compared to the lock components discussed earlier.

To return to Colt and his description of making the lock frame:

> The action of the machines commences by fixing the centre, and drilling and tapping the base for the arbor, which having been previously prepared – the helical groove cut in it, and the lower end screwed – is firmly fixed into position, furnishing a definite point from which all operations are performed and to which all the parts bear relation. [It passes through] twenty-two distinct operations [with] the majority of the tools used revolving cutters [i.e. processing by milling].

Other processes included: 'being introduced between hard steel clamps through which all the holes are drilled, bored and tapped for the various screws' i.e. the use of the drill jig. The part is then 'ready for finishing by hand, which consists in merely removing the rough edge or burr'. Colt's language reflects more modern engineering usage, 'operations' for example, and his approach highlights the need for reference planes and dimensions:

> All the various parts are made by machinery... and... travel independently through the manufactory, arriving at last, in an almost complete condition, in the hands of the finishing workmen, by whom they are assembled, from promiscuous heaps, and formed into firearms, requiring only the polishing and fitting demanded for ornament.

ENGLAND EMBRACES THE AMERICAN SYSTEM OF MANUFACTURES

Colt particularly identifies the value of simple dedicated machines and processes: 'It has been found advantageous to confine each one to its peculiar province, rather than to employ any more comprehensive machine, for several operations'.

Experiments[5] on original Colts suggest that they were not perhaps as interchangeable as the Colonel suggests – 'this being a typical Colt exaggeration.'[6] This is confirmed by the historical consultant, R.L. Wilson[7] of Colt: 'The parts in revolvers that need individual hand attention are the hand[8] and cylinder stop in particular. Sometimes the cylinder stop needs some filing'[9] – Springfield being more advanced in this respect. Contemporary illustrations of Colt's Hartford factory were made for the United States Magazine of 1857.[10] This was the model for his later London factory; one of the illustrations shows the assembly room. In this room parts were 'jointed', parts picked from a large supply to give a best fit, and each part received the weapon serial number: 'This was essential because the parts became separated during bluing, case hardening and engraving, and it ensured that the correct parts were subsequently re-united'.[11] This indicates limited interchangeability in the components of the weapons, which is also likely to be further impacted by any distortion on heat treatment.

Following the presentation of the paper, a vigorous discussion began. Mr Freedman introduced Mr Adams who exhibited 'a revolver of his own invention', the double acting Adams & Deane; again 'the barrel, the lock frame and top bar were all forged out of a bar of iron'. The forging required for the Adams & Deane was larger than that required for the Colt. A Mr May contentiously then began the focus on the manufacturing issues:

> He was of the opinion that the machinery used in the manufacture was not of the accurate description generally employed in this country, the tool marks were more evident than in the machine-made work produced at Manchester [Whitworth's factories were in Manchester] and no information had been given as to the price of production.

The vice-president Mr Rendell rephrased the question more gently and Colt gave some more detail including the information that 'Three hundred persons were employed and about one hundred arms finished per diem'.

In his paper Colt confirms our description of the British (and European) manufacturing approach above, writing the following:

> The manufacture of arms, both in Great Britain and on the Continent, is carried on almost entirely by manual labour, the various parts being forged, filed and ground into the requisite form by workmen in their own houses, the barrels alone being forged, bored, and ground in manufactories established for the purpose, and machinery being employed only for cutting out the stocks (blanks).

Colt complements Enfield: 'Under the intelligent direction of Mr Lovell, steps onward have been made in the use of machinery for some portions of the work'. The subsequent visits to the USA[12] confirmed: 'The system of forging gun barrels in this country by the trip hammer, worked by a water wheel or steam engine, is inferior to the English rolling mill as applied for the same purpose... but the adoption of the latter method was spoken of as probable'.

Returning to the meeting, Mr Hodge added to the discussion:

> Perhaps one of the most striking evidences of practical skill was the variety of crooked cutters[13] used for shaping... it would answer to the purpose of English manufacturers to send an intelligent foreman over to the States, to examine what was now being done there. The gun makers at Birmingham might certainly learn much from studying Colonel Colt's system of manufacturing, and their work would be produced at infinitely less cost, if they abandoned the system of employing a number of men, each working independently in his own ill-lighted, badly ventilated, and inconvenient workshop forming and filing up, without any uniformity, parts of arms, which would be produced in much larger quantities, in less time, and with perfect identity of form by the simple machines used by the Americans.

As we will see later, the USA was visited and it had dramatic effects on the Birmingham trade.[14] Colt was elected an Associate of the Institution and awarded its Telford medal in the following year.

Colt leased a building in 1851 on the bank of the Thames, formerly used by workmen constructing the Houses of Parliament, fitted it with a steam engine and began to produce revolvers for sale in January 1853. Revolvers were produced on the top three floors of the building. Planing machines and heavy lathes used to make new tools were in the basement.

A comparison of UK-located USA versus UK manufacturing technology applied to revolver making in the 1850s has been made,[15] including descriptions of Colt's London factory, by Dickens, and Adams & Deane's factory. These accounts indicate a much larger degree of mechanisation in the Colt factory. This included a 'universal' milling machine, a profile copying milling machine, a four head rifling machine, a drill clamp or jig and a special purpose screw machine[16] and tooling for repetition or fast changeover with 'Morse'-like drill tapers.[17] Colt's London factory included American forging machines – a Drop Hammer patented in 1853. There is some debate as to whether Gage Stickney an employee, or Colt the patentee, is the inventor.[18] There were also British die forging machines – a Ryder multiple station forging engine patented in 1841. Colt employed British labour, around 200 hands, with twenty-five to thirty American technicians. However, following disagreements with the establishment on the relative effectiveness of muzzle-loading revolvers and capping breech loaders by the cavalry and the arrival of peace in the Crimea there was less opportunity and need for Colt to manufacture in the UK. In 1858, to Colt's financial relief, the building was transferred to government control and became a Small Arms Repair Establishment and a training school for Armourer Sergeants.[19]

It is important for us to recall that Colt used Eli Whitney Junior as his contractor in 1847 to deliver his first successful military order – clearly he learnt much from him and his own factory manager Elisha Root. Root was one of the founding fathers of die forging and the creator of much of the tooling used in Colt's factory![20]

Visits to the United States

In the UK we were watching these changes in the USA with interest and following the exhibits of Colt and Robbins & Lawrence at 'The Great Exhibition of 1851'. The government selected a number of people to ostensibly visit 'The Exhibition of Industry in New York'. Particularly in the second set of visits they were to see a number of armouries including Colt's at Hartford, the Springfield Armoury and the Sharps, Remington and Robbins & Lawrence factories. In 1853 the visitors were Joseph Whitworth, Professor John Wilson of South Kensington and George Wallis. Whitworth was the master of precision and standardisation and along with James Nasmyth was a former employee of Henry Maudslay, the pre-eminent machine tool maker of the early nineteenth century.[21] In 1803 Maudslay had constructed, to the outline designs of Jeremy Bentham and Marc Brunel, the forty-five machines of the Portsmouth pulley block making machinery – the earliest large scale use of machines for mass production and some of the inspiration for Blanchard. Wallis was formerly headmaster of Birmingham School of Art[22] – and consequently the head of the only school of rifle design in Britain.[23] In 1854 the second mission was led by John Anderson, chief superintendent of the Woolwich Arsenal, subsequently the author of the paper on the application of copying machinery in woodworking, including **3.2**, inspired by his visit. He was officially accompanied by Lt-Col. Robert Burn and Capt. Picton Marlow, and perhaps unofficially by Joseph Greenwood[24] of whom we will see more later.

The visitors in the reports of their missions of 1854 and 1855 determined that parts from any Springfield Musket made between 1843 and 1853 (the Model 1842 Springfield percussion musket) would interchange. The reports of Whitworth and the main visiting committee are reproduced in *The American System of Manufactures* by Rosenburg: both verify true interchangeability. The committee noted the barrels and locks from 1844 to 1853 would interchange with one another and that: 'The lock fitter takes the parts of the locks from heaps of each promiscuously, and often making up forty or fifty without having to use a file'. Of particular interest is that Whitworth notes to the committee that muskets could be assembled in 3 minutes at Springfield. This musket was the first fully interchangeable product to be made in large quantities in both USA national armouries. It had been designed by Thomas Warner, using components from the short-lived Model 1840 flintlock musket. This had been designed by Benjamin Moor of Harpers Ferry explicitly to be capable of manufacture with interchangeable parts.

The Report of the Select Committee on Small Arms

The first mission report made up one of the significant pieces of evidence submitted to the Select Committee of Small Arms – which addressed 'the difficulty of procuring muskets' and 'the plan of a large government factory' for their production. The committee of fifteen included Birmingham industrialist and materials innovator George Frederick Muntz, and published its report on the 12 May 1854. The report itself has two volumes and over 500 pages – its summary is, however, straightforward.

The report begins with a statement of the current system for 'the production of muskets', stating that:

> The component parts are… procured by open contract and are after inspection and approval placed in the Ordnance Stores. There is a separate contract for each of the separate parts each is subjected to a strict view… accuracy of shape and size is secured by a system of gauges, while other tests… ascertain the strength of the work and the soundness of the material… A portion of the materials… is issued to the government establishment at Enfield, to be there set up, that is *fitted* together and finished into complete muskets. The chief part of the material is, however, delivered to contractors… for setting up… the arm… It is again subjected to a view.

Lord Raglan (recall the Charge of the Light Brigade, 1854) 'was not satisfied with the present mode in which that supply is obtained' and Sir Thomas Hastings, who gave the above evidence, said that: 'the experience of the last three years as a convincing proof of the impossibility of obtaining a sufficient supply of muskets under the present system' – the country was at war in the Crimea. There had been considerable delays in the procurement of 28,000 rifled muskets commenced in May 1851, contracts were entered into in February 1852 and the muskets were not delivered until November 1853. Similar issues had arisen in artillery carbines: only 500 of 2,000 required had been delivered, and in procuring 20,000 barrels for the new pattern (1853) musket – the tendering process had been very unsatisfactory. The trade had made many excuses: 'strikes… difficulty in procuring coal… illness of a skilled artisan… .an accident to machinery'. The Birmingham trade was not performing well. Col. Bonner who had superintended the supply of muskets to the East India Co. by the contractor system for twenty years however, commented that in spite of great fluctuations in demand, '12,000 in one year to 58,000 in another… the East India Co. had never found difficulty in procuring a supply nor had they ever found that the contractors took advantage of an emergency to raise their price'. However, the East India Co. had different standards for their weapons; the India Pattern Brown Bess of the Napoleonic Wars was initially created by diverting East India Co. Muskets to the Ordnance. There is much comment in the collector's literature of the differences in quality between true Land Pattern and India Pattern Brown Besses. The quality changes within the 1853 Pattern rifle musket would have been a further stretch for the East India Co.'s suppliers. This is the critical step that triggered the government factory: 'they have not succeeded in satisfactorily supplying the Board of Ordnance with the quantity of the superior class of arms required in the last three years for the use of the British Army'.

As a result of the change to the new pattern musket 'upwards of 900,000' would be required and the 'Board of Ordnance propose to take the whole manufacture into their own hands' and to plan 'an establishment capable of producing 500 muskets a day'. The reports of the visits to the USA described above and work by John Anderson, chief engineer at the Woolwich Arsenal, were pointing to the use of machinery. 'The advantages of producing muskets by machinery are said to be, cheapness in the manufacture, an exact similarity in the several parts, so that they may readily be interchanged and replaced and above all, the facility of rapidly producing muskets and of increasing or reducing the supply according the requirements of the time.' It was considered desirable to obtain from America 'such machines as they possess' because of the difficulty of 'originating' such machines. Unfortunately, Anderson had to admit that he was not: 'practically acquainted with the manufacture of muskets' and the Board of Ordnance: 'that they had never seen the musket to which they had referred'. This naturally led to the second mission to the USA.

The report also records that Anderson and his colleague, Lt Warlow of the Royal Artillery, had visited the Birmingham gun trade in March 1853 as part of a visit to 'some of the principal manufacturing towns'. This was not directed specifically to the manufacture of small arms but they were disappointed in what they saw! Anderson quoted from his report made in 1853 and expanded upon it in his evidence: '"We then visited a number of establishments engaged in military musket and bayonet work, all of which, however are in a low mechanical state, and at least fifty years behind most of the other branches of manufacturing industry which we have been examining." I may mention that the other branches here referred to are: the cotton, flax and woollen trades, engineering and machine making, the tool makers of Leeds and Manchester, steel pen and wood screw making of Birmingham. Those we were very much pleased with'. Needless to say the Birmingham gun trade was not pleased with this evidence!

Whitworth, in his evidence to the Select Committee stated: 'that with regard to an American musket, taking a number of the different parts of it, there is greater identity than there is in the rifles manufactured in this country'. He was also sceptical: 'he did not believe that the several parts of different muskets made by machinery would interchange; they might be adapted but they would fit badly'. Wallis however emphasised that at Springfield: 'the lock, the barrel and the furniture [are] put into a gun stock… indiscriminately.' Naysmth also gave evidence considering 'the systematic introduction of machinery into the gun trade to be feasible and highly desirable'. He had also supplied the Russian government with machines for making parts of muskets.

The report then goes on to describe the individual processes used to create the components of the weapon and the impact of machinery. Machinery is noted to be used in stock making in the USA 'but the Walnut used in this country is harder' in some parts of barrel making although 'The accuracy of the bore is tested by the eye and no mechanical contrivance has been invented to supersede this'.[25] In gun locks: 'It appears that machines have always been used in the manufacture of gun locks and if a large number of similar pattern were required the employment of machines might be advantageously increased'. These machines were only noted to be used at the factory of Braziers in Wolverhampton and we have no detail on them apart from some 'puff' from Brazier including his intent to buy a Nasymth steam hammer. It is said of the other suppliers: 'Nothing has been done beyond the old system of working by hand'. Perhaps these machines were similar to the simple hand-operated tumbler mills that we have seen earlier, rather than the 'self-acting' powered machines used in the USA. It was clear that machinery could be applied – and that 'if the gun trade could have confidence that a large supply of muskets would be purchased by the Ordnance in future years, the manufacturers themselves would be anxious to introduce machinery wherever it could be profit-

4.1 *A contemporary photograph of the Ames lock inletting machine. (By permission of the British Library)*

ably employed'. Whitworth had suggested that the new factory should not be as large as originally envisaged but that the 'government to have an establishment as perfect as could be made, to produce a limited number, and *to set an example to other gun-makers.*'

The report then turned to the contentious issue of quality in the UK: 'gauges and mechanical contrivances are used' but much discretion was left to the viewer. Viewing also inserted delay into the supply chain as did the use of only one pattern 'although there may be many contractors for the same article… only one pattern is sent, so that while it may be retained by one contractor, the others are necessarily delayed'.[26]

The outcome of this considerable discussion was that the Select Committee recommended that the system of contracting should continue but that:

> A manufactory of Small Arms under the Board of Ordnance should be tried to a limited extent. This manufactory would serve as an experiment of the advantage to be derived from the more extensive application of machinery, as a check on the price of contractors, and as a resource in time of emergency and it should be arranged with a view to its economical working.

4.2 *A contemporary photograph of the Robbins & Lawrence copy milling machine, tooled up to make trigger guards. (By permission of the British Library)*

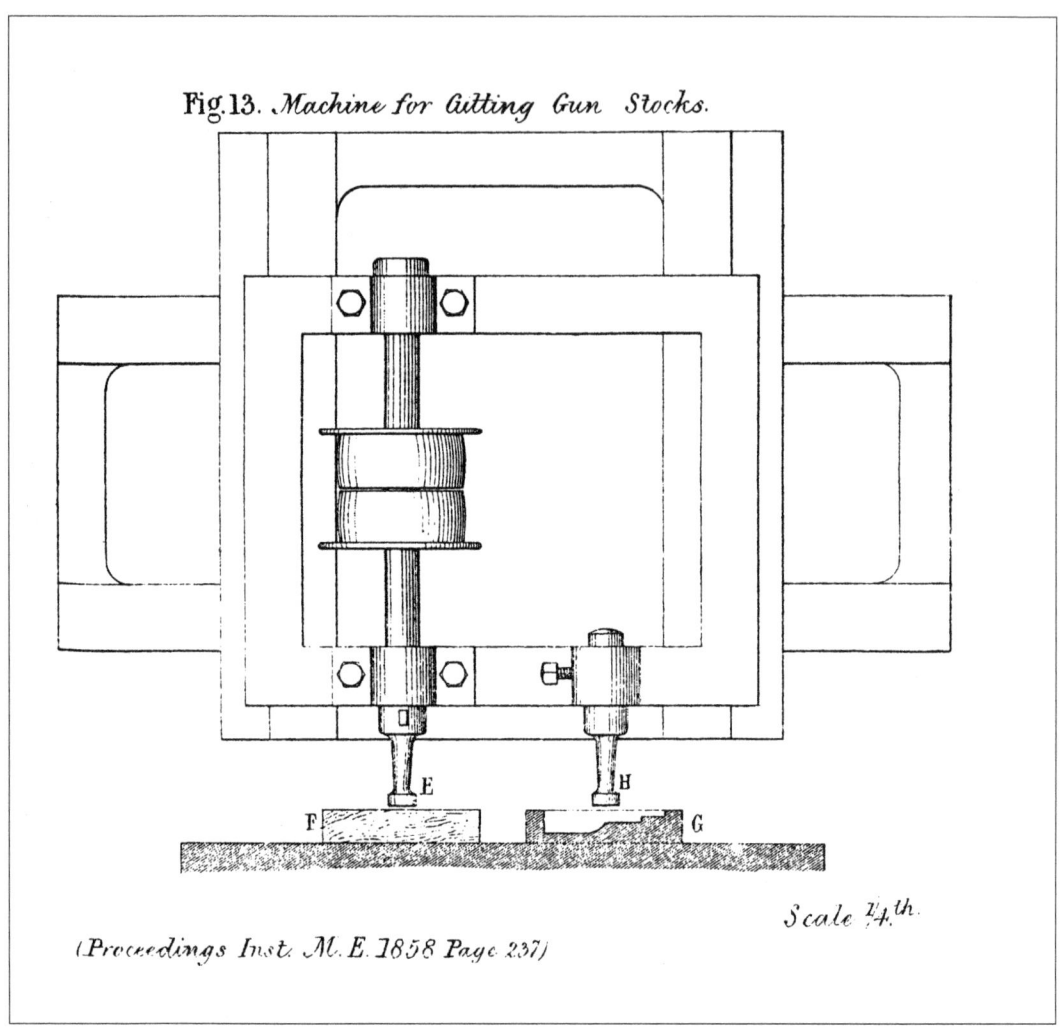

4.3 *Anderson's diagram of the principle of the inletting machine, from the Proceedings for 1858. (By permission of the President and Council of the Institution of Mechanical Engineers)*

The Committee considered that the new manufactory should be based on the existing site at Enfield.

There were two issues that the work had no answer for: would the application of machinery lead to cheaper weapons – it was not clear, and how would the smoothing of demand help the contractor system – perhaps the presence of Enfield would help?

Machinery Enters Enfield

Following the visits and report of the Select Commitee, the Americans tendered for machinery for making the Enfield rifle as follows: Ames of Chicopee Falls, Massachusetts to supply twenty-three Stock Making Machines costing $41,230 including delivery; Robbins & Lawrence of Windsor, Vermont to supply sixty-one milling, seventeen drilling and thirty-eight other machines costing $41,244, and gauges costing $5,600 from Ames. Further small wood, sheet metal and paper card working and cap making machinery was also tendered via Robbins & Lawrence. All tenders were subsequently accepted. Goodman 1866 tells us that £352,583 was spent at Enfield between 1 January 1854 and 31 March 1858, including £142,622 in wages.

There was therefore a major increase in wood and metal working machinery applied at Enfield. This included the Ames lock morticing machine shown in **4.1**[27] and other machines such as a specially tooled milling machine from Robbins & Lawrence shown in **4.2**, and the first mass UK application of the milling machine. **4.3** shows a schematic drawing by Anderson of the tooling for the inletting machine. The Ames machines were designed by Cyrus Buckland 'who was paid a fee of $1000 for his services'.[28] In 1859 the plant was described in a series of articles in the *Illustrated London News* in which it was stated the sixty-three components of the rifle were manufactured in 719 operations with an output of 1,200 rifles per week from 1,250 employees.[29]

This in turn led to the subsequent founding of Birmingham Small Arms and Whitworth's work to understand the mechanisms of rifling. Whitworth was not willing to work on manufacturing innovation for rifles before he believed in the technology of the product; he therefore began his work on rifling and was perhaps the first to take a scientific approach. When Whitworth began to build rifles he worked to tolerances of $1/2000$in; these rifles were tested in 1857.[30] Whitworth found errors of 0.03in in others' barrels!

As Colt has indicated, machinery had already crept into Enfield under Lovell's guidance, including a lock plate driller in 1842, percussioning machines in 1845, a Nasmyth steam hammer for large forgings in 1851 (invented in 1839[31]) and a Whitworth planer.[32] Other manufacturing machinery in military factories were nipple machines applied by Lovell in 1835, musket ball rolling machines by Napier between 1834–1842, mortice and loop machinery in 1851[33] and joining and percussioning and cock fitting machinery.[34] Within the civil trade Brazier's had 'machines' for lock making and steam machinery for rifling barrels. The new factory at Enfield 'for mass production' of the 1853 Pattern Enfield rifle was completed in 1859. James Burton, formerly of Harpers Ferry Armory and the Ames Manufacturing Co., had come over from the USA to take over leadership of production in October 1855.[35]

Because of its effective monopoly, the government had periodically attempted to reduce the impact of the Birmingham trade during the Napoleonic Wars. It created a lock and barrel making establishment at Lewisham from the old Armoury Mills in 1806. The site only survived a few years and was subsequently consolidated into Enfield in 1818, the move from Lewisham being favoured by the presence of water power there. Enfield was initially created in 1813 as a barrel making and inspection site that was capable of accommodating demand variation. Enfield subsequently had the capacity to make 26,000 barrels per year (compare this with the Birmingham output of more than 300,000 barrels alone per year between 1804 and 1815) with barrel boring and turning machines and rifling machines. It was explicitly decided by the government to *not* create this capacity, this the centre of the gun trade, Birmingham – because of the variable quality and problematic customer (government!) relations of the manual/craft-based industry. Enfield was an underdeveloped location relatively close to London. George Lovell was appointed as store keeper in 1816, gaining total control of the site in 1824. Later he was to control all procurement including that from the Birmingham trade. While Enfield grew, Lovell admitted that Birmingham remained: 'the great school in which the Royal Manufactory must always seek its hands' even though relocation could be problematic with workers returning to the

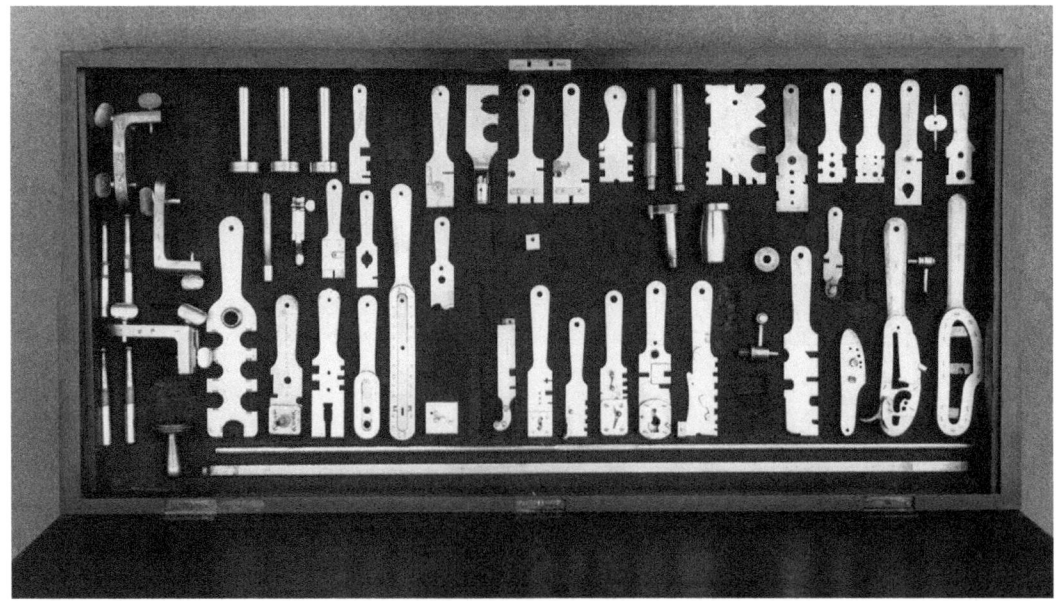

4.4 *Enfield gauges in their box. (By permission of Peter Jacques and the Muzzle Loaders Association of Great Britain – MLAGB)*

ENGLAND EMBRACES THE AMERICAN SYSTEM OF MANUFACTURES

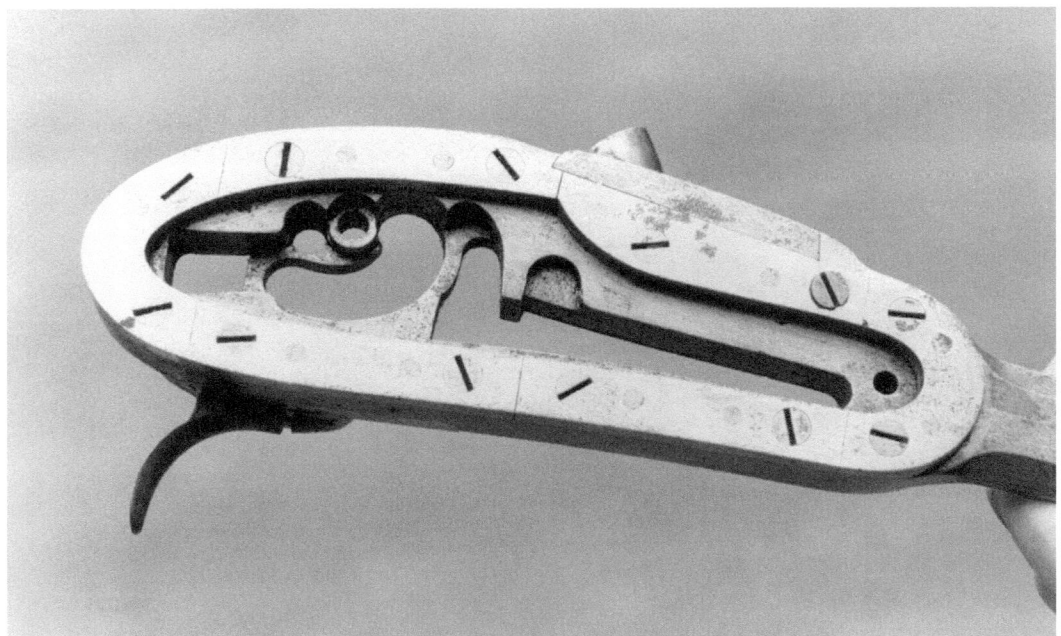

4.5 *An Enfield lock receiving gauge with a working trigger for testing assembly and function. (By permission of Peter Jacques and the Muzzle Loaders Association of Great Britain)*

sophistication and job mobility of Birmingham from the rural marshland of Enfield in a few weeks. Growth at Enfield and the employment of machinery was driven by the need to rearm with percussion weapons. Late in his career Lovell and his son had financial problems with their contractors, wrongly crediting some. Lovell was ordered to move to Birmingham to improve his control of the contractors and died in 1854, largely forgotten.[36]

The Birmingham gun maker and son of William Greener, W.W. Greener in the 1910 edition of the *Gun and its Development* summarises the story of the change for gun making in the UK:

> Mr Prosser was requested by the government to report as to the possibility of making guns to the interchangeable plan. This was in 1850; in 1852 Colonel Colt was examined by a committee of the House of Commons with reference to the same subject, and upon the strength of his representations a Commission visited the United States; the result of that visit was the [sic] founding of the Enfield Factory,[37] the purchase of American machinery and the introduction of the interchangeable system of manufacture into England at the close of the Crimean War.

Richard Bissell Prosser[38] did comment in his evidence to the Select Committee on John Jones who went from Birmingham to the Russian Armoury at Tula with others

THE BIRMINGHAM GUN TRADE

4.6 *The front of the gauge with a lock. (By permission of Peter Jacques and the Muzzle Loaders Association of Great Britain)*

4.7 *The rear of the gauge with a lock. (By permission of Peter Jacques and the Muzzle Loaders Association of Great Britain)*

4.8 *A simple tumbler gauge. (By permission of Peter Jacques and the Muzzle Loaders Association of Great Britain)*

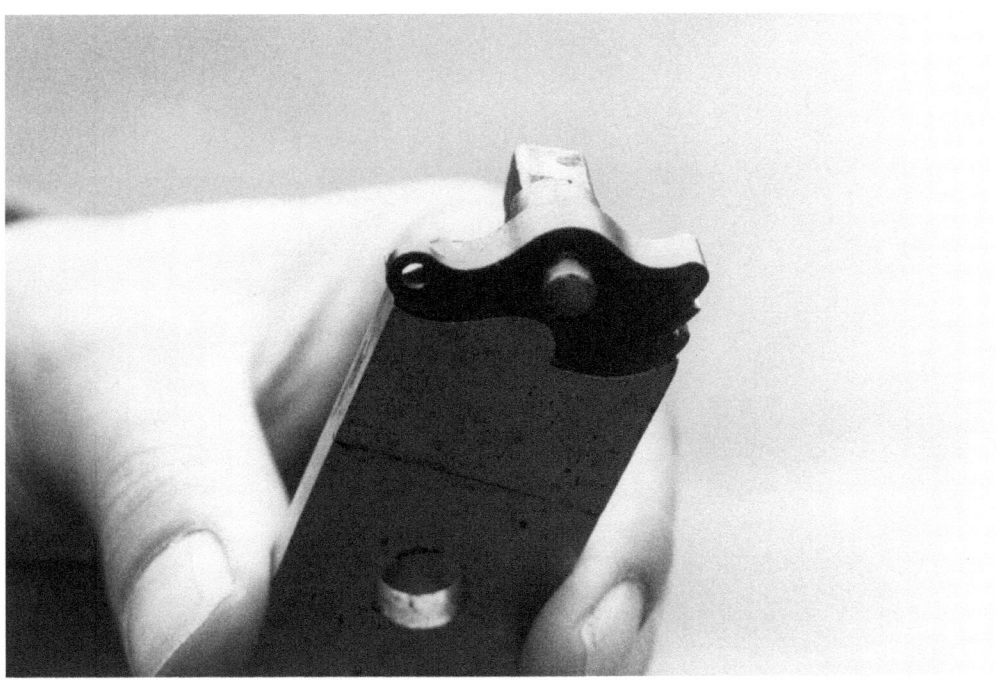

4.9 *The simple tumbler gauge checking form. (By permission of Peter Jacques and the Muzzle Loaders Association of Great Britain)*

4.10 *The simple tumbler gauge checking hole and pin pitch. (By permission of Peter Jacques and the Muzzle Loaders Association of Great Britain)*

4.11 *A more complex tumbler gauge checking form. (By permission of Peter Jacques and the Muzzle Loaders Association of Great Britain)*

4.12 *The complex gauge checking pitch. (By permission of Peter Jacques and the Muzzle Loaders Association of Great Britain)*

4.13 *The complex gauge checking a dimension. (By permission of Peter Jacques and the Muzzle Loaders Association of Great Britain)*

4.14 *The complex gauge checking a dimension. (By permission of Peter Jacques and the Muzzle Loaders Association of Great Britain)*

4.15 *The complex gauge checking a dimension. (By permission of Peter Jacques and the Muzzle Loaders Association of Great Britain)*

including his father, Richard Prosser, and was perhaps instrumental in introducing an interchangeable system there.

The Royal Small Arms Factory at Enfield, England was equipped with American-made machinery, particularly for wooden stock making and was by 1858 capable of manufacturing rifles to the Pattern of 1853 with completely interchangeable components.[39] Even William Greener was impressed: 'Enfield the seat of government manufacture of small arms, will become a celebrated place in future history; its productions being now one of the wonders of the present age'.[40] However, the American engineers making the machines were most critical that the gauges they were supplied with in 1851 for measurement of the British patterns were only accurate to $\frac{1}{64}$in.[41] Particularly important to this interchangeability were the USA-supplied[42] sets of 1853 Enfield gauges, shown in their original box in **4.4**, including receiving gauges. Jacques[43] shows excellent photographs of their use, some of which are reproduced here. **4.5** shows the receiving gauge for the assembled lock with a working trigger to allow functional testing. **4.6** and **4.7** show the gauge and a completed lock. **4.8** shows a simple tumbler receiving gauge, checking form, **4.9**, and pin and hole pitch, **4.10**. **4.11** shows a more complex tumbler receiving gauge checking form, pitch, **4.12**, and length, **4.13**, **4.14** and **4.15**.

In 1857 orders were also placed with the London Armoury Co. of Bermondsey to make interchangeable Pattern 1853 rifles with machines[44] in their factory in Victoria Pack Mills. This company had grown, with some complexity,[45] out of the Deane, Adams & Deane business with Robert Adams being required to devote his whole time and attention to the 'Manufacturing Department'. Adams subsequently went back into business on his own account in 1859. The company survived ten years and was wound up in 1867.[46] This was perhaps as a result of deciding to supply the Southern cause in the American Civil War or as a result of a financial crisis in the city consequent on honouring Southern debts. In 1863 the agents of the North had stopped buying arms from England because local makers could satisfy the federal demand, supplying the Confederacy meant continuing to manufacture! During this period the company focussed on rifle making and appears to have contracted some revolver making to Pryse and Redman or Calisher and Terry.

Robbins & Lawrence themselves were fatally a contract manufacturer for the 1853 Enfield during the desperate need for the arm in the Crimean War while the manufacturing capacity was being built in England. Many of their own machines came from Ames.[47] The company was contracted to supply 25,000 rifles with a promise of 300,000 more via the agents of the British government, Sir Charles Fox and John Henderson. Lawrence was cautious about the order because of the cost of tooling up. Beset by endless production problems, the company did not meet the contractual deadline defaulting in September 1856 with 14,600 rifles yet to be delivered and a debt to the agents of $73,000. With a penalty of $5.00 per rifle the company owed their creditors $146,000. The British government foreclosed and Robbins & Lawrence failed. Their machinery order for supply to

Great Britain was for $46,445.60.[48] This failure provided an opportunity for other machine makers.

The Ames Recessing Machine

The Ames recessing machine from the Enfield Armoury was preserved in the Science Museum, London in 1956 and is still on exhibition. It was described[49] by the Keeper of the Department of Mechanical and Civil Engineering as follows:

> The purpose of the machine is to cut out the recess in the stock to receive the lock. This is done in five operations by the application of cutters carried in a five sided frame which may be rotated about the central axis. Each cutter may be moved in slides vertically and horizontally by a handle. The stock is clamped on a table which may be moved at right angles to the other motions by means of an upright handle on the left, which acts through a quadrant on a rack under the table. Alongside the stock is a 'former', a block of hardened steel which has cut in it an exact copy of the recess required to be cut in the gun stock. Parallel to each cutter is fixed a tracer. This is a pin of the same dimensions as the cutter, which enters the former and is caused by compounding the motions to trace out all the complex recess that is to be excavated by the corresponding cutter. An interlock mechanism makes it necessary to return the driving belt to the loose pulley before the frame can be rotated to bring the next cutter into position. Air from a small fan mounted at the rear of the machine is directed through two tubes onto the stock and former in order to blow away the chips of wood.
>
> The machine ran at 5,000rpm and the five operations could be carried out on it in less than a minute.

The Ames company also supplied many plain milling machines to the gun making business from the mid-1840s.

Tolerances and Technology Mobility

Tolerances in gun making – *variations* – in the 1850s were discussed by Maj. Alfred Mordecai, a key figure in the USA Board of Ordnance until the Civil War, in his Report[50] of the Military Commission to Europe 1855–1856 to the USA Ordnance Board as follows:

> At the Arsenal of Vienna, the variation allowed in the diameter of the *bore* is half a point or 0.0038in. In the English arms it is said to be only 0.001in. In our armouries (USA) the allowance of variation was formerly 0.01in but in new arms it has been properly reduced to 0.0025in.

The UK 0.001in and USA 0.0025in tolerances confirm the levels of precision in use in one of the most precise dimensions. Goodman[51] confirms: 'Improvements have been made since that time (the American visits) in the direction of producing a higher standard of close fitting, as the English authorities were not satisfied with the American standard in this respect'. Whitworth and others were also roundly criticising the earlier practices of working to 'a bare sixteenth' of an inch or 'a full thirty second',[52] 'just under, or very bare'[53] or even 'a fine 1/64 of an inch'.[54]

Tolerances on the 1853 Enfield were +0.0015in.[55] Writing in 1880, Fitch gives us a later view and calibrates against that of the filing jigs of the turn of the nineteenth century: 'Uniformity then meant to within a thirty-second of an inch or more, where now it means within half a thousandth of an inch'. This shows the continuing evolution of the definition of interchangeable. Further, Mordecai reports in the text of his Commission when discussing the Enfield rifle:

> This musket has a barrel 39in long, with a bore of 0.577in; it is rifled with three grooves, which have a twist of 6½ft or half a turn in the length of the barrel; the grooves are 0.014in deep at the breech, diminishing to 0.004in (scarcely perceptible) at the muzzle. In the arms, made by contract, at Birmingham, the depth of the grooves is uniform; the contractors having no machinery adapted to cutting grooves of increasing depth.

Tolerances must be considered with respect to the technology to measure them. In 1855 Naysmith identified that Whitworth was measuring routinely to 0.0001in with his notional 'Millionth Machine'[56] and his difference gauges were being built in steps of '5,000th of an inch', *The Times*, 13 July, 1855. In North America the Vernier calliper was being applied, in 1851 Brown & Sharpe had sold four.[57] Increased measurement accuracy in the Vernier is based upon the comparison of two accurately divided scales. However, at Colt, when the Verniers were first put into service, enough discrepancies showed up in their standards for Colt to temporarily favour a certain amount of 'tailoring' in their fits.[58] As engineers will be aware, the application of the Vernier is still problematic for measurements of any precision but we should also recall some of the evidence that we have for 'jointing' or fitting in the Colt factory.

While the UK had led machine tool and materials technology in the early nineteenth century,[59] particularly with respect to precision, this leadership was passing to the USA in the 1850s.[60] Whitworth himself made observations to this effect.[61] Technology had and was however still going in both directions via materials and people.[62] Writing in 1826, Lovell notes that there were over fifty British artificers employed at Springfield and twenty-one at Harpers Ferry.[63] The UK retained materials leadership – Colt[64] and Springfield in 1855[65] purchased UK materials in the USA. Whitworth invented fluid pressed steel in 1856 to remove porosity and improve material properties for both ordnance and small arms barrels.[66] Bessemer sealed the production of cheap steel with the invention of his converter, also in 1856.[67] Barrel making innovations also went to the USA, barrel welding leaving in 1858 via William Onyans of Birmingham to arrive in Springfield in 1859 and subsequently in Remington of Illinois in 1868, and hook rifling technology transferred in 1861.[68] James T. Ames brought Onyans to the USA. When the barrel making equipment arrived in the USA it was used by the same operators – no skill change was required. The operators took a two-thirds reduction in their price per piece and their output went up fivefold! This gave them a wage increase and saved the government about 7c on a musket price of about $13.00.[69] From the above it will be clear that by 1859, on both sides of the Atlantic, truly interchangeable weapons – the M1842 Springfield and the 1853 Pattern Enfield – were being made to familiar levels of metal cutting precision with similar technologies. The two key technology steps were the use of sets of precision gauges and the accurate management of the edge dimensions of powered milling cutters.

Capital in the UK

There is much debate amongst economic historians[70] as to the reasons for the late adoption of wood and metal working tools for military gun making in the UK when compared to the United States. This debate tends to overlook the high capitalisation of barrel making in the UK compared to the USA: we must recall that a gun at this time was a lock, stock and *barrel*. Much of this debate centres on whether there would be sufficient steady government work to make the capital investment viable. It was clearly necessary in barrel making in the UK, but perhaps labour flexibility in the large metal manufacturing cluster[71] gave an alternative approach for other parts of the mechanism of military guns. It is likely that the total demand for military barrels created by the East India Co. demand and the demands of the African (formerly slave) trade for military style barrels combined with the physical demands and scale of the task, allowed the application of power driven machinery. There was a steady annual demand for 100,000 to 150,000 of the flintlock African trade guns, these being supplied with a beechwood stock stained black, brown or vermilion.[72] They were usually of variable quality – this is one of the reasons for the creation of the Proof House. Also, as Prosser[73] surveying patents in 1881 shows, Birmingham was highly innovative in all trades associated with the manufacture of tubes from the 1750s – a gun barrel being a particular variant of a tube. Echoing the Birmingham approach to innovation, Prosser also tells us that old musket barrels were used as gas pipes in the early experiments on gas illumination at Soho in the early nineteenth century!

5

THE BIRMINGHAM TRADE AND THE BRITISH MACHINE MAKERS RESPOND

The success of these new factories at Enfield and the technology within them caused much disquiet in the Birmingham gun-making industry. By 1858 the manufactory at Enfield was turning out 2,000 rifles a week. These were mainly the long Enfield rifle of the pattern decided in 1853 by a military committee. The committee had, among its advisors, Wilkinson and the Birmingham gun makers Greener and Westley Richards.[1] Westley Richards made one of the first prototype Enfields[2] and later introduced the 'monkey tail' capping breech loader sometime between 1858 and 1861.[3] The population of Birmingham was beginning to approach 200,000 from its population of 70,000 in 1800 and of 6,000 in 1700 – the military gun trade employed 7,000 at its peak in the 1850s.

A vigorous correspondence was also going on in *The Times*, challenging the competences of the Birmingham trade and highlighting Whitworth and other Mancunian manufacturing skills. This can be seen in the Joseph Whitworth and Co. Press Cuttings 1854–1862, held in the Library of the Institution of Mechanical Engineers. An article in *The Times* said on 13 July 1855: 'It is no slight affair that the skill of the mechanical engineer should at length be brought in aid of gun making' to reverse 'the arrest of progress' in the trade. A response in the *Birmingham Journal* on 21 July 'called upon the trade to look to their own interests' and was critical of the eulogised Whitworth, ridiculing the test barrels of '10 and 20in' he was building in his rifling experiments 'for an expenditure of £10,000'.[4] Goodman responded to *The Times* on the 31 July:

> To the charge brought against the manufacturers of Birmingham, that they are behind the times in the use of the mechanical aids that the present day so abundantly affords, I reply that Birmingham men are not often charged with neglect of their own interest… They know better than anyone can tell them, at what point to the use of machinery ceases to be profitable, and when hand labour must be resorted to.

He closes his letter:

> The Manchester men, in the small attempt they have made to aid in the supply of materials for rifles, thought they could dispense with hand labour. If I am not wrong, they have, at some pecuniary cost to themselves, discovered that they have committed a mistake. When I see our Town Hall transferred to their Market Street, or the ring of the anvil in our streets exchanged for the sound of the spindle, I will then begin to believe in the possibility of Manchester making guns or teaching Birmingham how to make them.

On 4 August the *Birmingham Journal* reported that on the previous Saturday, Messrs Cooper & Goodman had taken their workmen via the Great Western Railway to the Greswold Arms Inn, Knowle, for a day of sports and amusements followed by dinner for two hundred. Goodman gave a speech after the dinner echoing these themes and highlighted: 'You are not to forget the Enfield factory; and you are not to forget that unless you wish Enfield at some future time to carry away all your military work, you must be diligent now'. The address was warmly received, both 'operative' and 'employer' being 'alive to the serious nature of the ordeal through which the gun trade of Birmingham is passing'. In spite of the ordeal, as the *Journal* says: 'The remainder of the day was spent in a most agreeable manner'.

Westley Richards was a staunch supporter of Whitworth and his use of machinery.[5]

This chapter examines the use of machines in the Birmingham industry subsequent to the exchanges above and reviews the technology of cartridge making and rifling in the UK. It also discusses the final technical steps that transformed the special purpose machines and gauging techniques used in gun making to the core of the mainstream of mechanical manufacturing.

Birmingham Small Arms

To compete with the new arms factory, the larger firms in Birmingham 'subscribed capital towards the building of a factory to make rifles with machine tools… called the Birmingham Small Arms Co. (BSA)'.[6] This was founded on 7 June 1861 when eleven[7] of Birmingham's best known gun makers again gathered at the Stork Hotel to consider 'forming a company to make Guns by Machinery'. The shareholders were Joseph Bourne, Joseph Bentley, John Cook, Joseph Cooper, John Goodman, Issac Hollis, Charles Playfair, Pryse and Redman, Joseph Smith, John

5.1 *A copying or inletting machine still in use at Birmingham Small Arms in 1917. (By permission of the Birmingham Museums and Art Gallery)*

5.2 *A Blanchard-style copying lathe, still in use at Birmingham Small Arms in 1917. (By permission of the Birmingham Museums and Art Gallery)*

Swinburn, Thomas Lawden, William Tranter, Thomas Turner,[8] Thomas Wilson and, Benjamin and Henry Woodward. All of these men also maintained separate manufacturing facilities during the entire history of The Birmingham Small Arms Co. Ltd prior to its initial voluntary liquidation. Some of these were quite large, such as that of Bentley & Playfair.[9] The liquidation was due to issues in the manufacture of 40 million Berdan cartridge cases for Prussia. The business was subsequently reconstituted in 1873 as the Birmingham Small Arms and Metal Co. Ltd. John D. Goodman was Chairman. Goodman (1817–1900) was a Birmingham factor from about 1830 and, as well as his responsibilities in BSA, a partner in Cooper & Goodman. He was an active agitator for the improvement to proof introduced in the 1854 Gun Barrel Proof Act[10]. He gave evidence to the Select Committee of 1854[11] and had clearly learned from his participation, changing his views from the ones he had expressed earlier!

One of the first actions of the company was to order stock-making machinery from the Ames Manufacturing Co. in the USA and the bulk of machinery, including metal working machinery, from Greenwood[12] of Leeds and 'engines and shafting from Hick and son of Bolton'.[13] **5.1, 5.2, 5.3** and **5.4** show machines and gauges from 1917. **5.1**[14] and **5.2** show an Ames-like inletting machine (profiling the nose cap of a SMLE – Short Model Lee Enfield) and a Blanchard type lathe (cutting two SMLE stock components at the same time). **5.3** shows the final assembly of Lewis guns during the First World War and **5.4** gauges including receiving, plug and go, no go gauges.[15] Cary McFarland was

5.3 *The final assembly of Lewis guns at Birmingham Small Arms in 1917. (By permission of the Birmingham Museums and Art Gallery.*

an American appointed Works Engineer at a salary of £400, the new works including a Proof House and a room 'for the manufacture of the tooling and dies required in the establishment' being built on a 25 acre site in Golden Hillock Lane, Small Heath. 'Models and gauges were obtained from Enfield and full permission granted to inspect that establishment whenever information was required'.[16] Golden Hillock Lane was subsequently, and perhaps more appropriately, renamed Armoury Road! **5.5** shows the exterior of the factory, **5.6** a Victorian photograph of the factory interior and **5.7** the factory exhibition stand at the exhibition of 1862. BSA had 771 machines in place by 1871.[17]

Goodman[18] compares Enfield and BSA:

> At Enfield, every part of the gun from the earliest stage is produced in the factory. This has not been found necessary in Birmingham; advantage has been taken of having the source of supply so near at hand, to obtain certain parts in an unfinished stage from the manufacturers of the town... The factory has been planned on a scale to produce 1,000 guns per week.

The first gun to be made by BSA was a short Enfield rifle musket for Turkey. Its lock plate was marked with 'B.S.A. CO. 1865' and the Turkish cypher. It was presented to McFarland who was returning to the USA at the end of his two-year contract. By 1866 the company had recorded a profit of £7,000.[19] The first orders for British weapons

THE BIRMINGHAM GUN TRADE

5.4 *Gauges used at Birmingham Small Arms in 1917. (By permission of the Birmingham Museums and Art Gallery)*

5.5 *An early photograph of the Birmingham Small Arms factory. (By permission of the Birmingham Museums and Art Gallery)*

5.6 *One of the Birmingham Small Arms workshops in the 1860s. (By permission of the Birmingham Museums and Art Gallery)*

THE BIRMINGHAM GUN TRADE

5.7 (Left) *The Birmingham Small Arms stand at the exhibition of 1862. (By permission of the Birmingham Museums and Art Gallery)*

5.8 (Below) *A Mark III Snider cavalry carbine breech.*

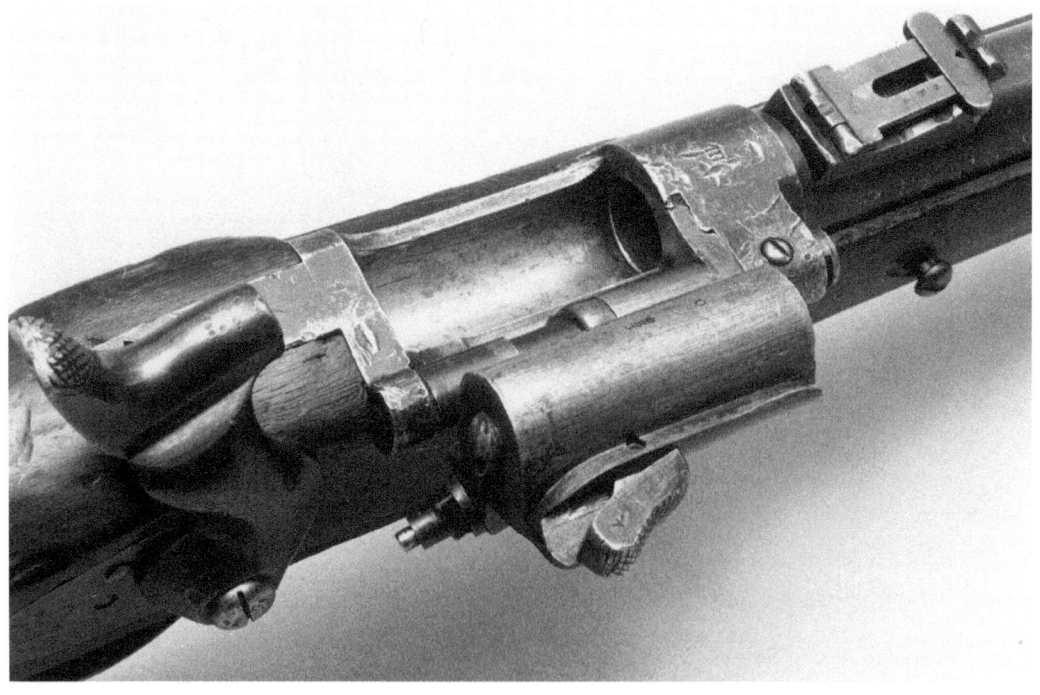

placed on BSA were for the Snider breech-loading conversion of the muzzle-loading Pattern 1853 Enfield. By the end of 1871, BSA had converted about 156,000 Sniders and made 52,475 new guns. **5.8** shows a Snider carbine and its breech. W.W. Greener[20] contrasts the price of machine-made Sniders at £2 10*s* with those 'made by hand, with triangle bayonet, £2 5*s*' showing the greater than 10 per cent premium associated with manufacture with machine. The purpose-designed single shot breech-loading Martini Henry was adopted in 1871 but BSA had to wait until the 1880s for significant orders of the new weapon. Orders for the new bolt action rifles followed in 1890. During the 1870s BSA was the largest privately owned rifle manufactury in Europe. In the 1890s, Mauser in Germany and the Osterreichische Waffenfabrik Gesellschaft in Austria became larger. While BSA primarily made weapons for the Empire, these two factories competed for the wider world market.

The London Armoury Co. was also a competitor but disappeared as a rival in 1865, it was replaced, under new management, by the London Small Arms Co. in 1867. This company and BSA formed an agreement to regulate competition between them.[21]

Other Birmingham Factories with Machines

Other competition from BSA came from the National Arms and Ammunition Co. Ltd (NAACo) which was formed in 1872 to make 150,000 Mauser rifles for the Prussians. NAACo also held the Martini patents. The prime movers in NAACo were Thomas Greenwood, John Batley and Westley Richards who had come together in 1870 to form The Westley Richards Arms and Ammunition Co., subsequently NAACo. They purchased land in Montgomery Street, Sparkbrook and acquired the assets and liabilities of Ludlow's Cartridge Works, Belmont Street, which was re-erected near Perry Barr and named the 'Holford Mills'. The first Managing Director of NAACo was Maj.-Gen. William Dixon who had previously been the superintendent of the Royal Small Arms Factory. He joined BSA as Secretary in 1875. The movement of Dixon and the involvement of Greenwood in many of these enterprises clearly show the characteristics we now expect of manufacturing clusters. NAACo was purchased by the government following its wind up in 1882.[22] The government also had a rifle repair – no doubt mostly a manual process of fitting spares – factory in Sparkbrook in 1887.[23] This was purchased by BSA in 1906 when they acquired it as part of the Peddie Small Arms Co. It was immediately liquidated.[24]

Holford Mills was occupied by the Gatling Arms and Ammunition Co. Ltd,[25] led by James Accles. Accles, an Australian, was apprenticed in Colt's Manufactory at Hartford in 1867. He was subsequently employed by the Russian government on the Berdan rifle and its ammunition. In 1872 he came to the UK and established a factory at Ward End, Birmingham, to make caps for the Berdan. Between 1873 and 1886 Accles was employed by the Gatling Gun Co. In 1873 he established factories for arms and ammunition in Canton and Shanghai, China, returning to England in 1875. He occupied Holford Mills in 1880. The name of the business subsequently changed to Grenfell & Accles, then to Accles Ltd and then to Accles & Pollock in 1901.[26]

In his *History of the Birmingham Proof House* Harris notes:

> The rifle gauges used in connection with the Prussian contract were bought by the firm of W.W. Greener and remained in their storeroom until 1914, when the firm acquired the old government factory, The Tower, in Bagot Street, and employing these identical gauges manufactured Mauser rifles for the Belgian government.

Oliver Greener[27] said when he took over the factory in December 1914: 'There was an inch of dust on the floors, the factory having laid empty for years, but by the end of January 1915 the tool room was up and running and the factory made its first Mauser by August of the same year'. This shows us both the way a cluster operates, the life of machine tools and the value of gauges! **5.9**, **5.10** and **5.11** show the Tower factory at this period when it was known as the National Rifle Factory No.2.

5.9 *The Tower, Bagot Street. Barrel shop, Birmingham Proof House Library. (By permission of the Guardians of the Proof House)*

5.10 *The Tower, Bagot Street. Barrel drilling, Birmingham Proof House Library. (By permission of the Guardians of the Proof House.*

5.11 *The Tower, Bagot Street. This shows Belgian Mauser rifle barrels, Birmingham Proof House Library. (By permission of the Guardians of the Proof House)*

The single approach of these factory system-based companies with their mechanised processes contrasts with the difficulty of handling early orders placed in Birmingham to produce the 1853 Pattern rifle musket, which required barrels from seven contractors to be assembled with locks from seven further subcontractors with minor parts from further contractors. The Birmingham trade had also struck for higher wages during the critical period of the Crimean War, thus interrupting supply; there was much to recover from in the face of this new competition. It is worth noting that the Ames machines are automated wood-working machines for stock inletting by copying. These consequently do not require the level of accuracy of metal-working machines for mechanism components. Whitworth was becoming active in this sector and would have no doubt influenced the levels of precision sought in England.

Thomas Greenwood

Thomas Greenwood was born in 1814. He joined Sir Peter Fairburn as Chief Draughtsman of his Wellington Foundry, Leeds, in 1843. This followed service with Messrs Whitham and an earlier failed activity with his brother, making attachments for textile machines, that began in 1833.[28] While working with Fairburn, Greenwood met John Batley and both were taken into partnership in the Wellington Foundry. In 1856, the two dissolved their partnership with Fairburn and created Greenwood & Batley at the Albion Foundry, East Street, Leeds.[29] In his book on Joseph Whitworth, Atkinson tells us that Greenwood was one of the visitors to the USA.[30]

From their second site, the Albion Works, in Armley, Leeds, the firm supplied many wood working and metal working machines.[31] These included forging, outside turning, boring, lapping and rifling machines for barrels[32] as early as 1857, for the gun making – their most profitable line – and other businesses including sewing machines for the boot and shoe industry. Batley seems to have quickly moved into the background. Greenwood had strong links with the Swiss, who have always been strong in precision manufacture. Small arms ammunition -making machinery became a strength in 1862, as did the manufacture of ammunition itself in 1868 with the manufacture of solid cases for the Spencer cartridge. Other successes included the suite of machinery, a total of 948 machines including rifling machines, screw drop hammers, boring machines and lathes supplied to Russia in the early 1870s to build the small bore sliding bolt variant of the Berdan breech-loading rifle and its bayonet. The first trap door Berdan rifles were built by Colt in his Hartford factory. The machinery shipped to Russia was first installed in BSA. James Burton installed the machinery in Russia and it was in full production in 1873.[33] The Russians commented adversely on the quality of the Birmingham-made weapons when comparing them to Colt.[34]

Greenwood gave an number of presentations at the Institution of Mechanical Engineers, describing his own solutions for stock making[35, 36] – **5.12** and **5.13** from the papers show drawings of Greenwood's evolutions of the Blanchard lathe and the Ames inletting machine, including its tooling and bearing arrangements, **5.14** – and building the Boxer Cartridge.[37] Comparison of the machine in **5.1** to the drawing in **5.13** confirms that the machine at BSA was built by Greenwood. It is also instructive to note the similarities between the Ames inletting machine shown in **4.1** and Greenwood's machine. Greenwood died in 1873.

The early UK machine makers grew but were not usually located in Birmingham; they came from London, Manchester or Leeds as did Greenwood & Batley. However:

> Due to the failure of Robbins & Lawrence, who had supplied most of the American machinery [to Enfield], and to preoccupation of American manufacturers with the Civil War, it fell to Greenwood & Batley to spread the initial equipment for the [American] system to many continental armouries including Tula.[38]

The British machine makers certainly learnt from and exploited the American system.

Cartridge Making

Following the development of the Enfield muzzle-loading rifle, Britain was struggling with the steps necessary to adopt breech-loading firearms. The capability to make precise cartridges was critical to this. On 18 September 1866, the Snider conversion of the Pattern 1853 rifle with the Boxer centre fire metallic case cartridge (**5.15**) was adopted as the standard weapon for the British Infantry. One of the most significant features of the Pattern 1853 was the increase in sophistication of the manufacturing techniques used to make the later versions of rifle with precise interchangeable components. These manufacturing improvements, however, pale when compared with the production engineering needed to make the Boxer cartridge in the volumes required for the army of the British Empire – 1¼ million a week! The significance of the challenge and the merit of the response can be judged by the appearance in the Proceedings of the Institution of Mechanical Engineers for 1868, a paper by Thomas Greenwood describing the machinery for the manufacture of the Boxer cartridges.[39] The paper and discussion run to twenty-nine pages and forty-three plates! This section tries to capture the flavour of the problems addressed by Greenwood and his colleagues.

The cartridge was developed by Col. Edward Boxer, superintendent of the Royal Laboratory. Boxer 'improved' upon the cartridge designs of Francois Schneider and the London gun maker George Daw to add a coiled case that expanded on firing. Again, this idea had been anticipated by Joseph Needham.[40]

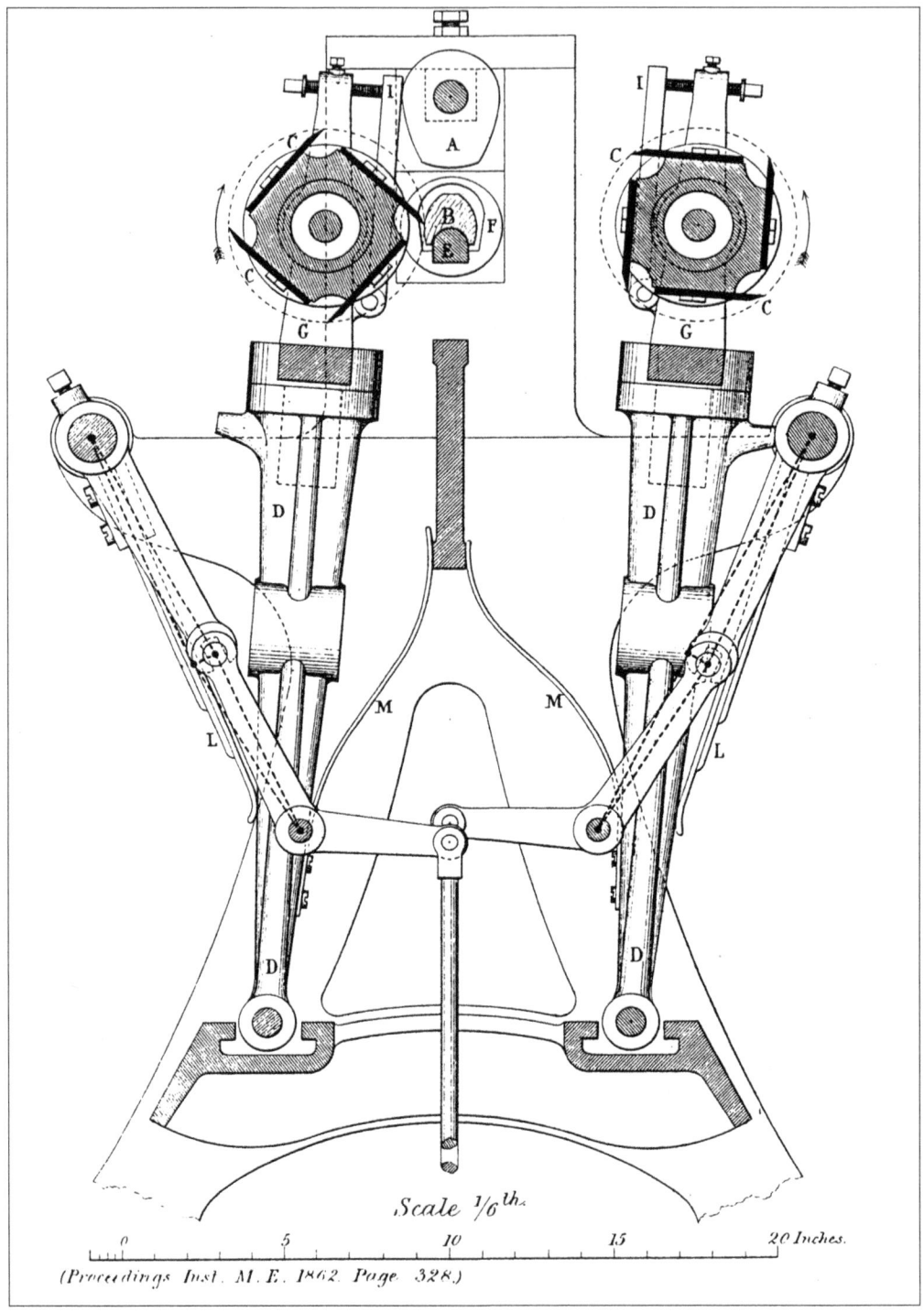

5.12 *Greenwood's variant of the Blanchard lathe, from the Proceedings for 1868. (By permission of the President and Council of the Institution of Mechanical Engineers)*

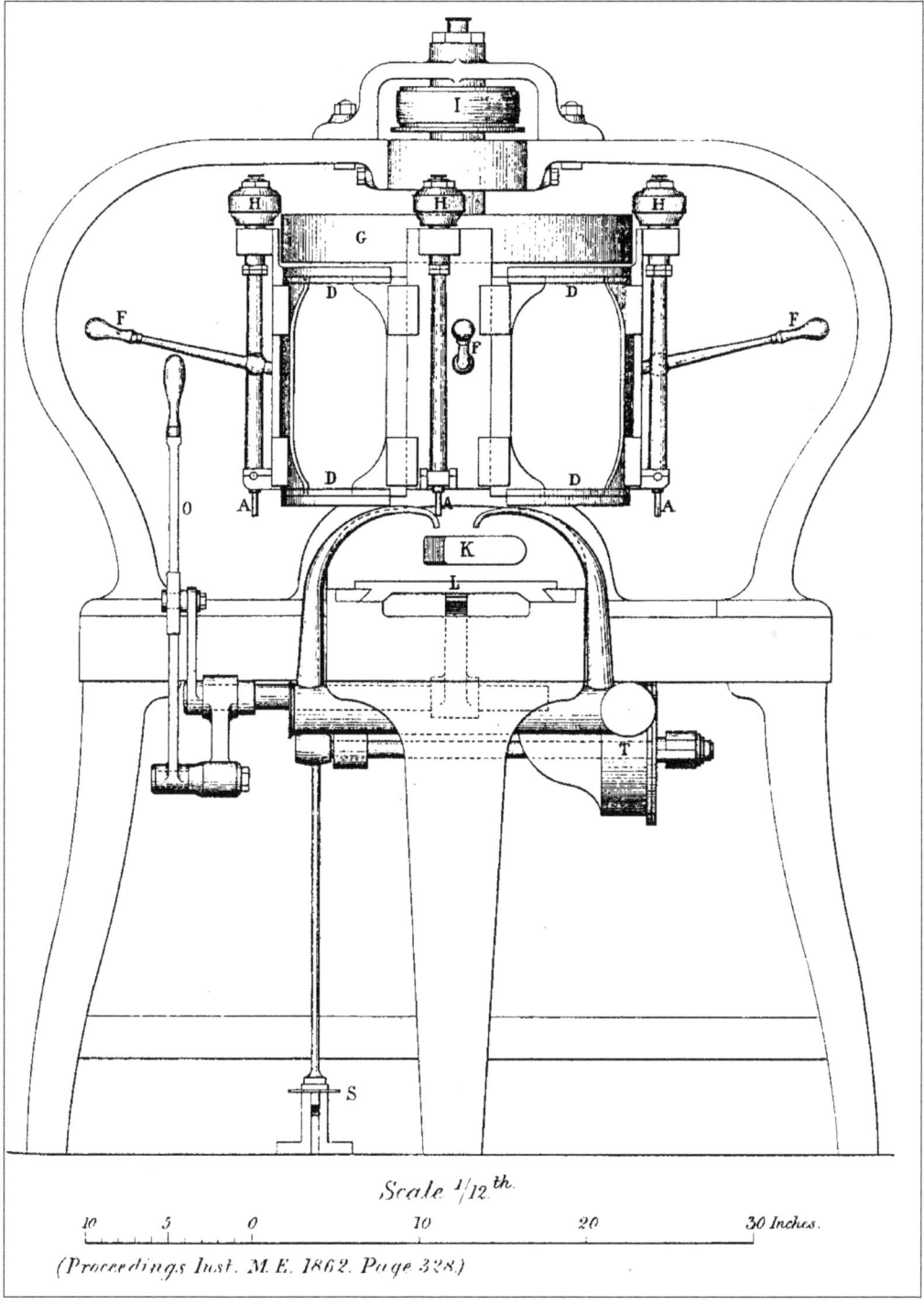

5.13 *Greenwood's copy of the Ames inletting machine, from the Proceedings for 1868.*
(By permission of the President and Council of the Institution of Mechanical Engineers)

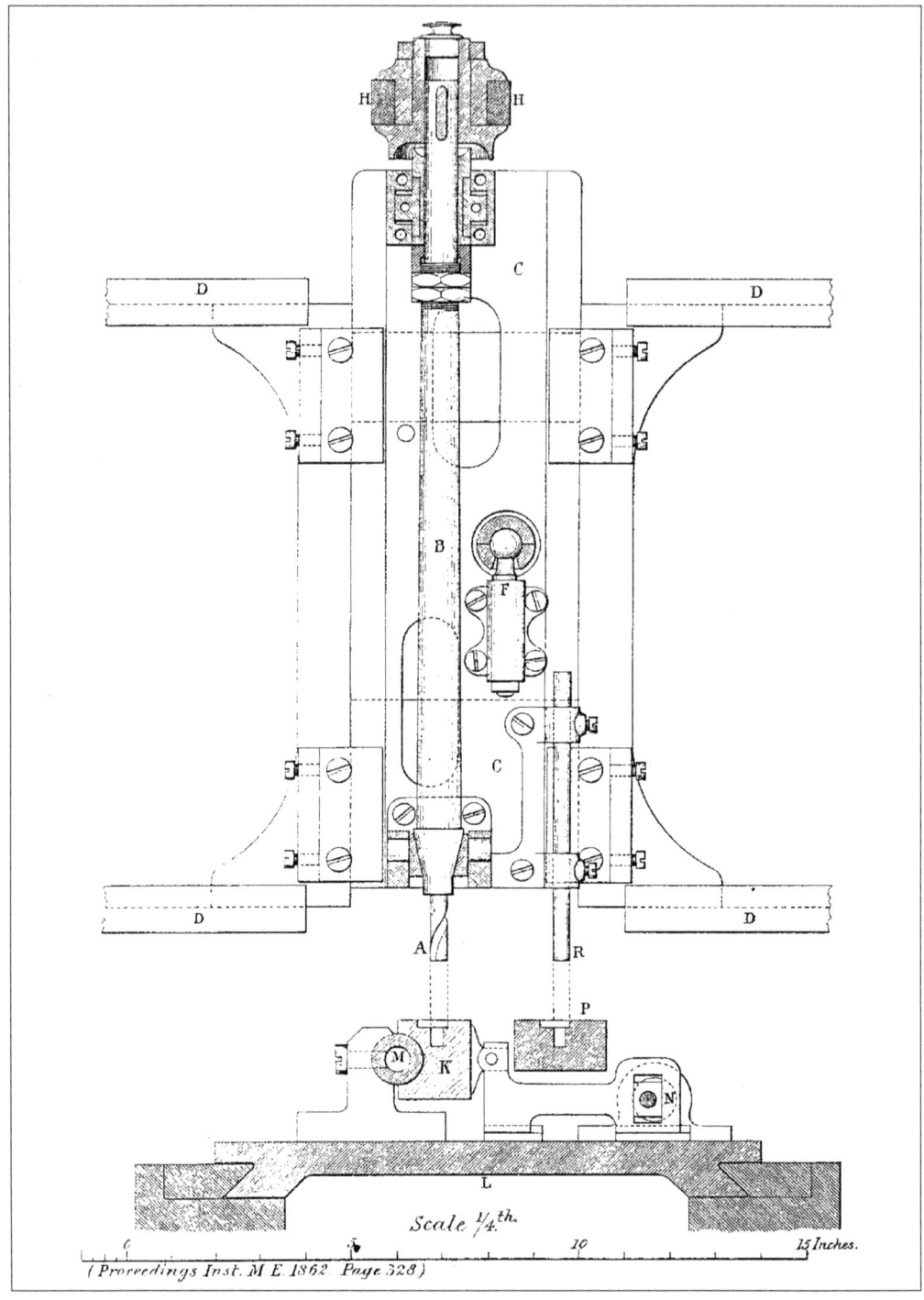

5.14 *Greenwood's inletting machine tooling and bearing arrangement, from the Proceedings for 1868. (By permission of the President and Council of the Institution of Mechanical Engineers)*

5.15 *A Boxer cartridge.*

Greenwood worked with Mr Davidson of Woolwich Arsenal to perfect the machinery. In total, twenty-one different kinds of complex machine had to be designed and constructed in order to manufacture and assemble the cartridge and its components. The output of the machines varied from 3,000 to 150,000 operations in a day of 10 hours. Many of the manufacturing processes used were similar to those that had been developed in order to manufacture the percussion cap – see for example the base cup-making machine in **5.16**. Many cartridge makers began as cap makers. The cartridge case is essentially an assembly of thin sheet metal components made by blanking, deep drawing (cupping) and coiling processes.

People – frequently orphan children – were very busy in the factory. Two key operations were carried on hand lever presses: one fixed the cartridge together and one fixed the bullet. These two 'machines' – simple tools to assist the operator, locate components and put them together – were expected to run at 3,000 per day or one completed assembly operation by the operator every 12 seconds. In order to achieve the output, Davidson would have needed seventy people on each of these operations. One of the plates shows the plan of the machine for gauging (sizing) the bullet and pressing in the clay base plug – including three rather feminine pairs of hands (see **5.18**.) The machine runs at 40,000 a day. Each of those pairs of hands is therefore feeding a component every second for 10 hours![41] **5.17** shows the machine for making the clay plug.

THE BIRMINGHAM GUN TRADE

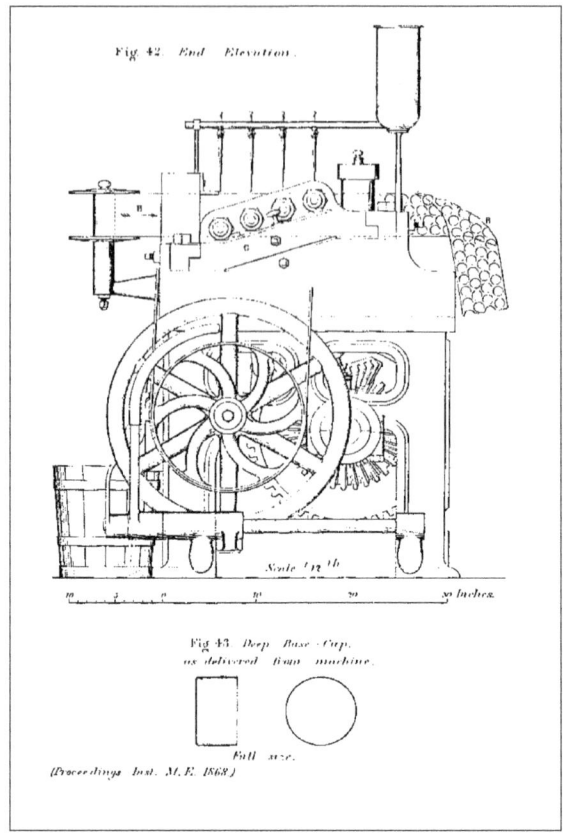

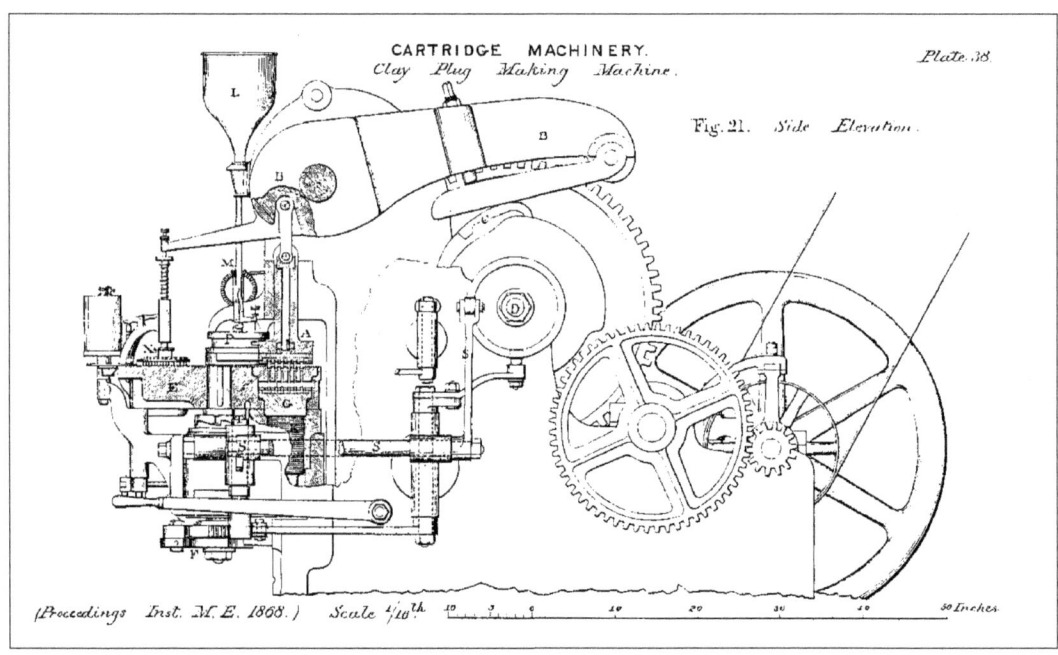

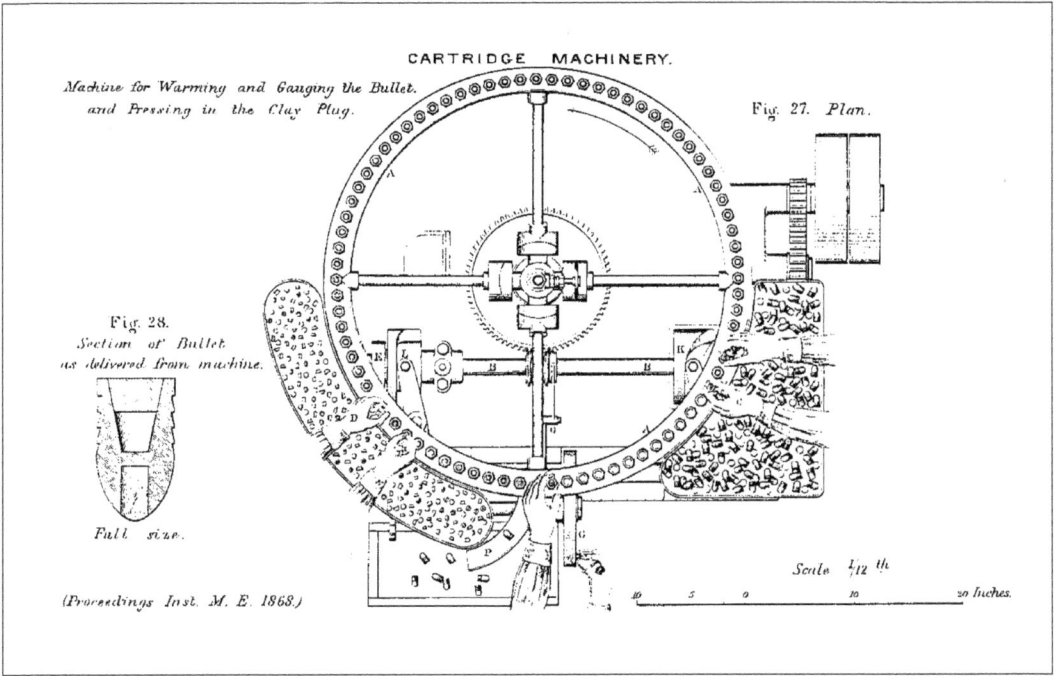

5.18 *Greenwood's Cartridge machinery machine for warming and gauging the bullet and pressing in the clay plug, from the Proceedings for 1868. (By permission of the President and Council of the Institution of Mechanical Engineers.*

Opposite Page

5.16 (Above) *Greenwood's Cartridge machinery base cap making machine, from the Proceedings for 1868. (By permission of the President and Council of the Institution of Mechanical Engineers)*

5.17 (Below) *Greenwood's Cartridge machinery clay plug making machine, from the Proceedings for 1868. (By permission of the President and Council of the Institution of Mechanical Engineers)*

THE BIRMINGHAM GUN TRADE

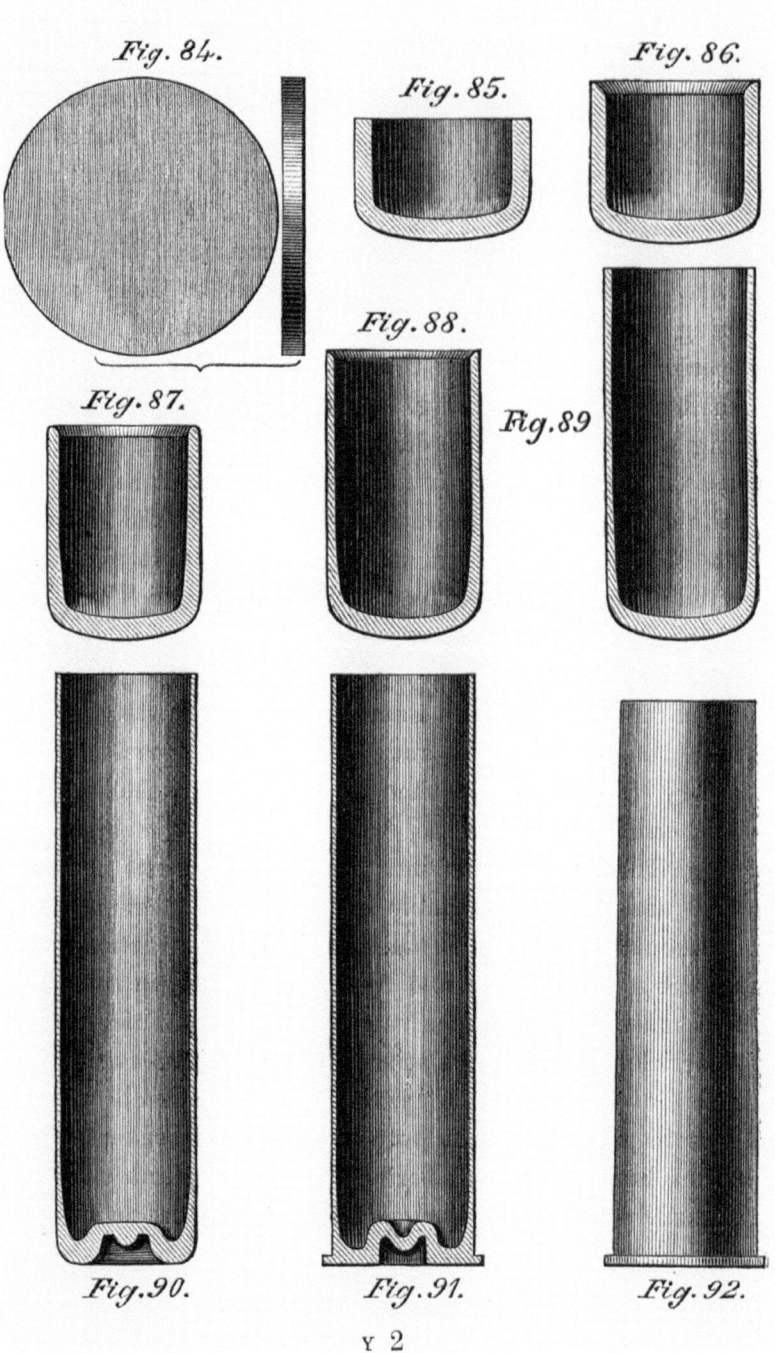

5.19 *The steps in deep-drawing a cartridge case. (Walsh 1884)*

In the discussion of his paper, Greenwood identified that the most serious difficulty in the development of the cartridge had been to make the rear of the cartridge sufficiently thick to resist the explosion and to also have a rim for extraction. After unsuccessfully attempting to make a single component to fulfil both needs, the developers created a five-component solution with nested thin walled cups assembled to a thick iron disc. The cupping or deep drawing process in turn gave them problems with puckering – folding – 'and several months experimenting' showed that this could be removed by making an 'exact' fit between the tools, an 'error of 1-$\frac{1}{1000}$ of an inch in the diameter was sufficient to spoil the cups turned out'.

The severity of the problems associated with developing early cartridges that were capable of resisting the pressures at the breech, even with black powder, can be inferred from cross sections of later black powder cartridges. In these, the material at the base is much thicker than that at the mouth. These cartridges are made by a different manufacturing process to that of the Boxer – multiple deep drawing (with no thickness change of the material) and wall ironing (with wall thickness change) steps – **5.19** shows this schematically. The material at the mouth was considerably worked in these processes and further worked as it was crimped to grasp the bullet, leaving the material highly stressed. This led to the phenomenon known as 'season cracking'. Cartridge cases in the tropics were found to very quickly crack on the onset of the wet season, especially near stables. There were very significant problems between 1906 and 1910, and in 1924. It was later established that for the copper rich (70 per cent plus) brasses used in cartridge cases, the presence of ammonia (from the decomposing horse urine) in a moist atmosphere could cause cracking or crack propagation over night in materials with significant residual stress. This leads to cartridges that might burst, 'putting off the marksman' as one senior soldier put it! The residual stress arose from the tooling used in the manufacturing process and the initiation of cracks in the cases during storage aggravated by the presence of sulphuric acid. The key paper,[42] even today, on the metallurgy of season cracking, from the Royal Laboratory, was written in 1921 and runs to 117 pages with its discussion! Its companion paper,[43] by F.S. Grimston – Deputy Director of Ordnance Factories and Manufacture, Simla India – identifies the manufacturing process that is at fault.

Returning to the Boxer cartridge, the trickiest problem in the development of the machinery to make the bullet was achieving the weight accurately. The weight of lead in the bullet was 530 grains 'and the margin for error was only 1½ grains above and below that, or 3 grains total variation'. The solution was to make the slug used for the bullet slightly oversize and extrude the excess 'through six small holes made laterally around the outer edge of the mould, which acted like safety valves'.

The most dangerous process in the factory was carried out on the capping machine. This is:

> The last of the machines employed in the manufacture of the cartridges and performs the final operation of inserting and pushing home the detonating cap to the anvil in the cap chamber. As this has to be done after the cartridge is loaded with the powder and ball and the cap charged with fulminate, it is necessarily a dangerous process; until recently it was performed by hand, but in consequence of a fatal accident the machine... was invented, which to a very great extent obviates all risk.

The capping machine invented ran at 20,000 a day but still had two operators to feed it. A rough calculation, however, says that if the old capping process ran at the usual assisted operator assembly speed of 3,000 a day, Greenwood and Davidson removed five people from the danger zone.

Even on its introduction, the Snider was recognised as a transitional arm. It was replaced when the Martini Henry was approved in 1874. The Martini Henry cartridges were very similar in construction to those for the Snider and many regiments retained the Snider until at least 1880[44] – no doubt the machines and people remained busy for many years!

The Birmingham Trade and its Slow Uptake of Machines

In spite of the protestations of Goodman, the Birmingham trade did not always perceive the important technical trends. In the discussion of Greenwood's 1862 paper, the Chairman asks if the Birmingham manufacturers have purchased any of the machines and Greenwood responds: 'At present Birmingham gun makers are paying a high price to London' for precision stocks. Greenwood's stock-making line cost '£8,000 for 500 a week capacity', allowing one to be assembled every 6 minutes. This position clearly changed when BSA later bought Greenwood's machines on the formation of the company. Things were also changing in lock making. Goodman said in 1874:[45]

> Till within the last few years locks were entirely the production of hand labour, the several parts were forged on the anvil by men whose wonderful skill became proverbial. They were afterwards put together by filers, to be finished by the polisher and hardener. At the present time the steam hammer and stamp are superseding the forge, and milling machinery is doing much of the filers work, but in no case, even when machinery is carried to the highest perfection, can the filer be dispensed with; the locks cannot be put together until all the limbs have passed through his hands to receive the final adjustment.

This echoes practice in the United States!

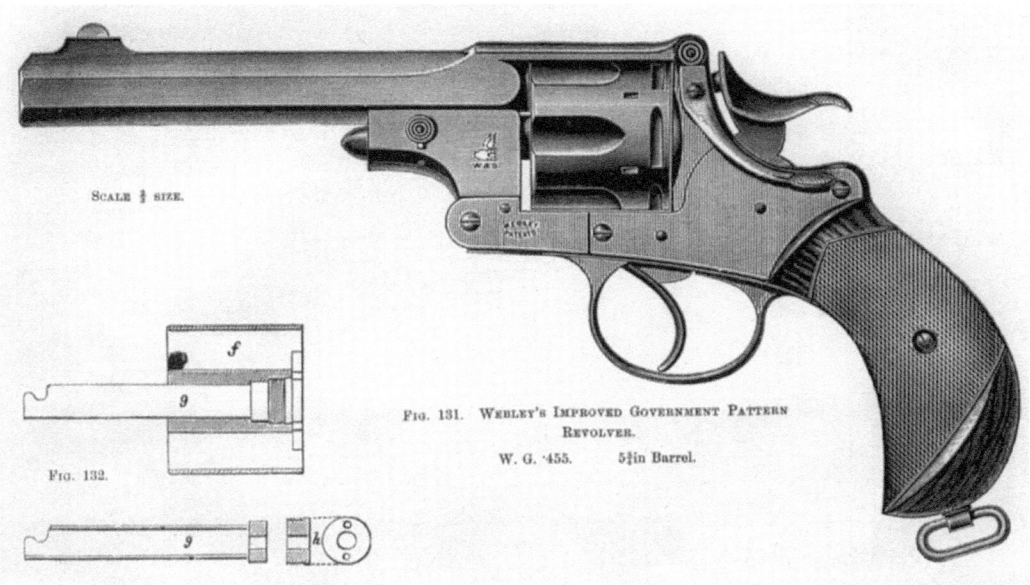

5.20 *A Webley revolver. (Walsh 1884)*

Webley and Revolver Making

The Birmingham gun trade made both longarms – rifles, muskets and shot guns – and revolvers. Webley was one of the major Birmingham revolver makers. In 1874, Dr Langford describes[46] the process of drifting or broaching and the milling of a forged revolver frame. **5.20** shows a characteristic Webley revolver. From a report on the 'mysteries' of revolver making at Webley's Weaman Street factory published in *Iron*:

> The pistol body… is a rough bit of malleable iron of the form and shape required. At the centre… is a square hole… to receive the cylinder. The first process is to force a long piece of cutting steel, called a drift, through this… and thereby cut it into shape. Two drifts, the second finer… than the first are used and the force required… sometimes amounts to 10 tons. This forms a standard to which all other parts of the work have to be done… The next operation is a very beautiful bit of machine work. The piece of iron is put on a block the size of the drift-hole, and by moving it backwards and forwards the sides are planed or milled quite true. This machine also cuts the recess by which the cartridge passes into the chambers… The milled body is next placed in a jig, in which all the action holes are drilled; then the strap is made for the handle of the pistol, and it is next cut along the top of the strap and round the body and the slot for the steel spring is cut in.

5.21 *A cast-iron boat as used in the heat treatment of actions.*

5.22 *Hand finishing Webley revolvers 'in the white', courtesy of the MOD pattern room.*

This description is extended in a further Victorian account,[47] which notes that the factory produces 400 revolvers a week and covers shotgun manufacture. The second account emphasises how vertically integrated the Webley business is: 'All the processes connected with gun manufacture with the exception of the milling of the tubes are carried out on the premises… The firm evidently make it a point to be as little dependent upon others as possible, but although they make admirable guns, they have to obtain the iron from others'. The visitor is however a little critical of the layout of the factory: 'The manufactory, we may mention at the outset, is somewhat divided, but this by no means interferes either with their completeness or the convenience of the artisans'. The second account describes copying or 'edge milling' at Webley to create components:

> These are placed on mandrils &c., corresponding to their internal configuration, and which, firmly held there, have their external contour delineated by machines fitted with tools which operate as a planing machine horizontally, as a slide-rest perpendicularly, or as rapidly revolving 'grinders' i.e. toothed steel wheels: the amount of 'cut' is determined by the 'tracer' which follows the contour of the model of the part desired to be copied and reproduced. The 'tracer' defines the tool cuts, the depth of cut is determined by the model traced, but all the machines, however numerous, are simply copying, though copying with undeviating accuracy, even to the three-thousandth part of an inch.

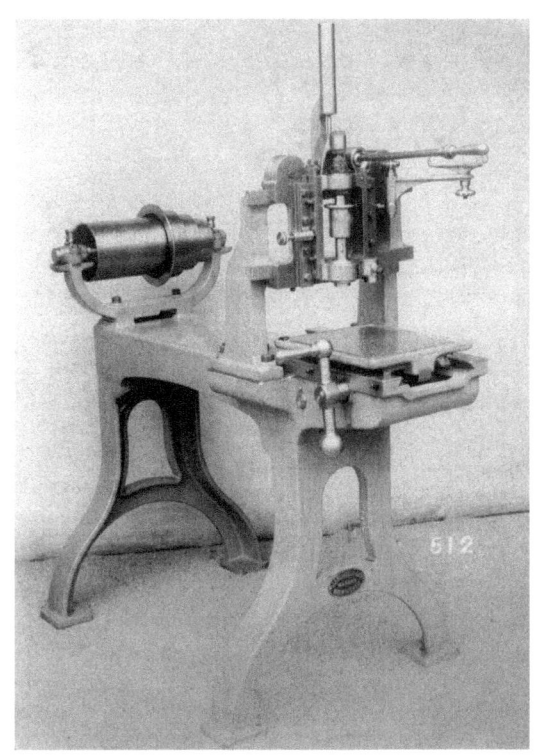

5.23 *An Archdale edge milling or profiling machine cost £50, Birmingham Proof House Library. (By permission of the Guardians of the Proof House)*

5.24 *An Archdale milling machine for magazine rifles cost £45, Birmingham Proof House Library, (By permission of the Guardians of the Proof House)*

5.25 (Right) *Archdale bullet making machine 1890 capable of producing 30,000 a day, cost £80, and with a coiled lead feeder it was £10 extra, Birmingham Proof House Library. (By permission of the Guardians of the Proof House)*

5.26 (Below) *Greenwood & Batley heading machine for cartridge making 1892, Archdale Agents, Birmingham Proof House Library. (By permission of the Guardians of the Proof House)*

5.27 Greenwood & Batley varnishing machine for cartridge making 1892, Archdale Agents, Birmingham Proof House Library, (By permission of the Guardians of the Proof House)

The anonymous author also describes the action hardening process – one of the critical steps to creating the distinctive finish of a quality firearm:

> An iron pan, about 10in by 4in, open at the top is partly filled with bone-dust.[48] The action work being completely embedded in the osseous particles, is put into a furnace, where it remains for about an hour, at the expiration of which period it is removed and turned out into a vessel containing cold water. This operation has the effect of thoroughly hardening the surface and imparting the beautiful colours that are some much admired for the tone and tint shown on this portion of the gun.

5.21 shows a hardening pan. Hand finishing was required for the final fitting of the revolver components together 'in the white' before heat treatment. **5.22** shows this process and the use of a hooded smoke lamp.

When Webley went to seek North American machines for his revolver factory in 1885 with £10,000 of capital to spend, he returned disappointed and purchased local machines.[49] There was significant use of Greenwood & Batley machines, also James Archdale of Birmingham was selling his own gun-making edging and milling machines and cartridge-making machines and acting as agent for Greenwood & Batley. His factory was founded in 1868.[50] **5.23**, **5.24** and **5.25** show Archdale's

SILVER MEDAL	GOLD MEDAL	GOLD MEDAL
1878.	1882.	1884.

KYNOCH AND CO.,
WITTON, near   BIRMINGHAM

By Appointment

Ammunition Manufacturers to   His Spanish Majesty King Alfonso XII.

MANUFACTURERS OF
SPORTING CARTRIDGES
IN ALL BORES: PAPER AND SOLID BRASS.

Military, Express, Rook Rifle, and Revolver Cartridges of every kind.
Punt Gun Cases, Gun Waddings, Percussion Caps, Anvils, and Re-loading Tools.

PATENTEES AND SOLE MANUFACTURERS OF THE
"PERFECT" RELOADING
METALLIC CARTRIDGE
(CENTRAL-FIRE AND PIN-FIRE),
THE CHEAPEST AND BEST YET INTRODUCED.

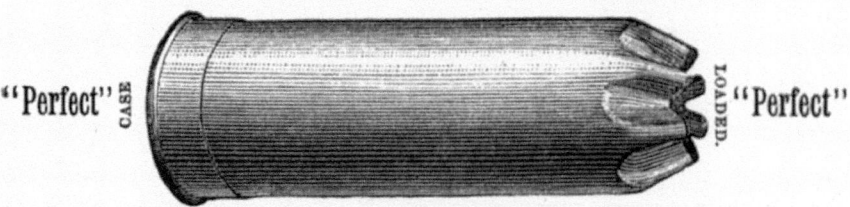

"Perfect" CASE   "Perfect" LOADED.

4, 8, 10, 12, 14, 16, 18, 20, 24, 28, ·410, ·360 GAUGE.

TESTIMONIALS AND SAMPLES ON APPLICATION.   PRICE LIST AND ILLUSTRATED CATALOGUE TO TRADE ONLY.

5.28 (Opposite) *A Kynoch advertisement showing their location as Witton, near Birmingham. (Walsh 1884)*

5.29 *A draw press in use at Birmingham Small Arms in 1917. (By permission of the Birmingham Museums and Art Gallery)*

5.30 *A large Archdale cartridge draw press, Birmingham Proof House Library. (By permission of the Guardians of the Proof House)*

edge milling and 'Magazine Rifle' milling and bullet making machines, **5.26** and **5.27** show later Greenwood & Batley cartridge machines for heading and varnishing respectively.[51] Note the similarity of the machine in **5.24** to the Robbins & Lawrence machine in **4.2**. Archdale was a Greenwood & Batley apprentice and by 1896 was employing 450 people – and was only taught to read and write in his fifties (between 1889 and 1899)![52]

Webley, the largest revolver maker, remained in Weaman Street near St Mary's until 1959 and then relocated to more modern premises in Park Lane, Handsworth in 1960. They continue to make air weapons today. Tranter, originally from Oldbury, also made significant numbers of revolvers of innovative design at Aston Cross, Birmingham, the business being leased to Kynochs, subsequently IMI, and then passing to The Aston Arms Co. For a time, Tranter worked in partnership with the Hollis Brothers. The Hollis family concentrated on the manufacture of longarms for military and export markets and became one of the largest Birmingham 'gun factories'. Tranter may have learned a considerable amount from the manufacturing and business approaches from his time with the Hollis's.[53] Tranter[54] 'made a considerable proportion of British revolving arms' but this reflects a 'modest market' rather than the 'large scale of his operation.'

Kynoch's, famous as a ammunition maker, began by making percussion caps.[55] A Kynoch advertisement of the 1880s is shown in **5.28**. The challenges associated with drawing processes related to ammunition making are indicated by the scale of the unguarded line shafting driven crank press at BSA (shown in **5.29**) drawing components of Lewis gun magazines from blanks in 1917 and the large Archdale cartridge drawing press shown in **5.30**. This machine, built in 1890 is: 'capable of drawing cases up to 8in dia – 7ft long. Weight about 24 tons. Price £800. They are now driven by large friction clutches instead of toothed clutch shown.'

Writing on Birmingham and the Industrial Revolution in 1974, Chambers observes:

> It was not until the last quarter of the nineteenth century that machinery was devised (or imported from America) which would enable standardised factory production to compete with the products of the myriad small masters who supplied the highly specialised products of the finished metal trades: light arms, jewellery, sporting guns, japanned products, etc.

The 1853 Enfield first brought that change into UK gun making and radically changed the form of an industry cluster.

> The Industrial Revolution was really a century long transition to mechanisation that had by-passed much of the gun making industry… Though a hard won living could be made from production of sporting guns, no two of which were identical or whose parts were not required to interchange, it was quite another

matter to produce many thousands of guns whose acceptance depended on interchangeability of parts. Very few of the 'manufacturers' in mid-nineteenth century Birmingham, though their combined output was immense, had the production capacity necessary for large scale production and even the largest made extensive use of subcontractors.[56]

These developments, and the end of the Civil War, caused the craft trade to rapidly contract and caused the location of manufacture to move locally within Birmingham towards Small Heath and the outer ring of later industries, and nationally from Birmingham to Enfield.[57]

Pressures on the Industry

In the book *The causes of decay in a British Industry* published in 1907, two industry leaders, Charles and William Wellington (W.W.) Greener, famously bemoan these changes under the pseudonyms Artifex and Optifex: 'If Birmingham gun makers, possessing as they do un-rivalled knowledge of their craft, with the best skilled and most highly trained artisans at their disposal in factories equipped with the best machine tools, cannot find enough work to do, the fault lies not them, but with either Birmingham or the state.'

Birmingham and the State had upset the Greeners by imposing legislation, adding to the burden of the manufacturer by imposing taxes and restricting personal liberty, by making the fees at the Proof House and the Patent Office too high, by favouring the middle man rather that the manufacturer, by free-trade policies and by not supporting the British in overseas markets. The government was also populated by 'members of a certain social class… who spurned trade, commerce and industry'. Foreign governments had imposed tariffs to close markets – we also had much to learn from the Japanese about the way their government supported the manufacturer 'until the Japanese factory is able to produce similar goods at lower cost', in 1907! The German approach of scientifically based technical education however was criticised. They did nevertheless note the need for practical training: a training school had been started in the Proof House in 1901. On the 'middle men and dealers', the Greeners are harsh:

> It is just as easy to become an expert buyer of finished goods. Instead of a huge factory, *entailing the thousand and one terrors of manufacturing*; exposing the owner to labour strikes, heavy assessments, and keen competition, there is a compact warehouse or shop, a little unskilled labour, a few book-keepers and salesmen; less risk, less worry, easier direction, and greater profits.

The road developments in the central city of Joseph Chamberlain's Corporation Street also came in for criticism. Prior to 1880, Birmingham decided on an improvement scheme to give a new large thoroughfare and 'sweep away' many insanitary dwellings. The firearms trade was in the area to be affected, but the 'trade' was not actually dislocated until ten to twenty years later.

> The industry was carried on in leasehold properties clustered around St Mary's Square, and the lease of these properties lapsed between 1890 and 1905, when such of the land as was scheduled for the improvement scheme was acquired by the corporation, and the properties were demolished... From the point of view of the town the policy was not only justifiable, but wise and profitable...

The death rate in the redeveloped areas dropped from 53.2 per 1,000 during 1873–75 to 21.3 per 1,000 in 1879–81.[58] As the Greeners say, this required the building of new factories and workshops for the outworkers and threatened new factories built by the trade before 1880, thus damaging both key aspects of the trade – the large manufacturers and the independent in rented workshops. These developments are echoed by the development of the Queensway Inner Ring Road in the twentieth century which further ate away at the St Mary's area. Many of the photographs in this book were taken in the trade between these two great municipal developments.

In the 1910 edition of his book, *The Gun and its Development*, W.W. Greener notes: 'The military branch of the (Birmingham) trade was checked by the establishment of firearms manufactories at Lewisham and Enfield... and it is upon the sporting trade that the Birmingham industry still depends'. With the introduction of the factory system using machine tools to Birmingham gun making, a mechanised industry was added to the existing hand work industry. While much of the volume, mechanised true firearms industry has disappeared, there is still a sustained Birmingham trade in sporting guns – high added value products usually tailored to individuals, with considerable 'hand' working.

It is interesting to reflect on the performance of other gun-making locations. Much of the inspiration for interchangeablity came from France and Germany. What progress was being made in continental Europe? Liège was the other major centre of military gun making at this time. The Chairman of BSA observed that the English had the advantage of the Belgians in 'the more extended use of machinery, the use of which in Liège is discouraged by the cheap rate at which hand labour can be maintained'.[59] However, data on the number of barrel proofs at Birmingham and Liège[60] shows that Liège production overtook that of Birmingham in around 1875. It has been suggested[61] that the presence of cheap hand labour in Belgium, France and Austria caused them to adopt forging and milling machinery later than the UK, these two technologies being particularly important in the manufacture of one piece revolver frame and barrel components.

A feeling for how we were doing in Birmingham in around 1890 can be inferred from the observation that 110 of 800 in the Birmingham military trade at that time were filers, compared with 11 of 1,600 in Springfield.[62] We had moved on but there was still much to do – and the American author of the comparison may have had a somewhat rose-coloured picture of reality at Springfield!

It is useful, however, to distinguish the practices used in military gun making from those used in the USA private sector. Focussing on revolver making Howard[63] identifies that the private sector uses fitting or selective assembly (Smith and Wesson) in order to achieve both fit and function and argues that the military market 'had no cost imperatives'.

> The private sector of the American arms industry made weapons whose parts were not interchangeable [as a result of]... the economics of production, the inherent nature of machine tools, the functional requirements of the weapons, and the needs of the buyer. The fitting method of assembling arms was simply a more economical system of production than that of producing interchangeable parts.

1980 practice shows that: 'Only Remington claimed some models of shotguns were completely interchangeable' following a design review in the 1940s aimed at eliminating bench fitting. Winchester claimed 95 per cent interchangeability and Marlin, Harrington and Richardson, Browning, Ithaca, Colt, Smith and Wesson and Iver Johnson confirmed this pattern of non-interchangeability: 'An emerging trend is that new product lines are designed with ease in manufacturing in mind and hence achieve a greater degree of interchangeability' and 'Hand fitting in the higher-priced products is still deemed a virtue'.

Rifling

Perhaps the most particular process used in gun making is that of creating rifling. This process, unlike milling, has not found wide application outside gun making and for that reason is not particularly well understood. I am, therefore, including a description of the evolution of rifling processes. We should note as well[64] that there was no revision of the rifling techniques used in the UK when American machinery for other processes was introduced. Because of its military and commercial significance, it is difficult to find good accounts of the techniques of rifling.[65]

There are two technical problems in the creation of rifled barrels: the generation of the helix and the cutting process used to create the rifling grooves. Early methods to generate the helix used a good or master barrel as a rifling guide. In this case, a lead follower is used within the barrel to guide the cutter bar. Short lead replicas cast into the cut rifling have frequently been used to carry polishing abrasive to give the barrel its finishing touches. In North America a large diameter wooden bar with helical slots passing through a matching nut was used to form the rifling guide for the cutter bar. The wooden helix methods date from 1790–1800.[66]

The process is described by Andrew Ure in 1839:

> After the barrel is bored and rendered truly cylindrical it is fixed upon the rifling machine. This instrument is formed upon a single plank of wood 7ft long, to which is fitted a tube about an inch in diameter, with spiral grooves deeply cut internally through its whole length, and to this a circular plate is attached about 5in diameter, accurately divided in concentric circles into 5 to 16 parts, and supported by two rings made fast to the plank, in which rings it revolves. An arm connected with the dividing graduated plate and pierced with holes, through which a pin is passed, regulates the change of the tube in giving the desired number of grooves. An iron rod, with a moveable handle at the one end, and a steel cutter in the other passes through the above rifling tube. This rod is covered with a core of lead one foot long. The barrel is firmly fixed by two rings on the plank standing in a straight line on the tube. The rod is now drawn repeatedly thorough the barrel from end to end until the cutter has formed one groove to the proper depth. The pin is shifted to another hole in the dividing plate and the operation of grooving is repeated till the whole number of riflings is completed. The barrel is next taken out of the machine, and finished. This is done by casting upon the end of a small iron rod a core of lead, which when besmeared with a mixture of fine emery and oil, is drawn, for a considerable time, by the workmen from one end of the barrel to the other, till the inner surface has become finely polished.

Scloppetaria: *On considerations on the nature and use of rifled barrel guns, etc.* of 1808 includes a description of an alternative rifling process, highlighting that the barrel has to be accurately bored (from a tube in 1808), the rifling cut and then finished. The core of creating the rifling is to generate and cut the helix – Scloppetaria describes how the machines did this. A twisted square metal bar is created 'with great care… by filing and polishing… this in turn passes through two square guides… [called puppets in the later literature] to therefore generate the helix. This twisting bar guides the rifling cutter or file held in turn on a wooden bar'.

In his 1841 book *Engines of War*, Wilkinson describes in full a similar machine to that in Scloppetaria, with a multi-point cutter mounted on a wooden bar. He particularly describes the way the cut is put on: 'The cutter is elevated a little, by taking it out of the groove in the wood and placing underneath one or two slips of writing paper soaked in oil'. Wilkinson also describes a more modern machine that dispenses with the twisted rod. It uses:

> …a carriage freely traversing a rail-road and in the middle of this carriage is the dividing plate and rifling rod, the latter being made to turn by means of a pinion wheel as the carriage is pushed backwards and forwards. This machine is now generally employed for best rifles, and the old plan for military ones.

THE BIRMINGHAM TRADE AND THE BRITISH MACHINE MAKERS RESPOND

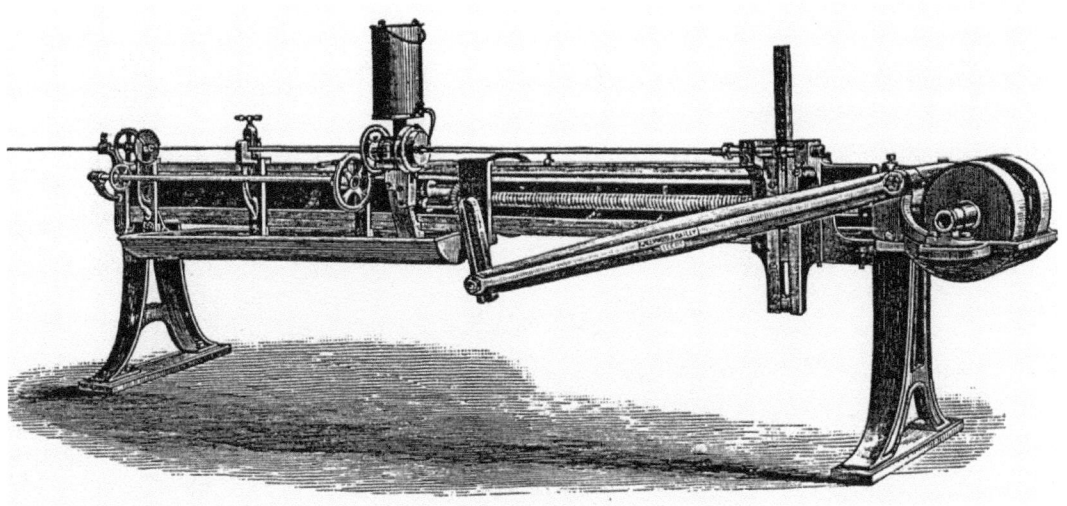

5.31 *A Victorian Greenwood & Batley rifling machine. (By permission of the Muzzle Loaders Association of Great Britain (Walsh 1884))*

This discussion was echoed by Lovell at Enfield. He reported to the Board of Ordnance in 1845[67] that rifling was done: 'by means of steel spiral rods filed out by hand, each degree of twist being given by its peculiar Rod, and the formation of these rods being dependant on the dexterity of one particular artificer who is growing old and his sight failing'. As a result of this, Lovell bought a machine from Evans and Sons of Wardour Street for £110: 'in which the spiral being given by an inclined plane, acting upon a rack or double toothed pinion, any twist may be given to the groove by simple adjustment'. Writing in 1881, Prosser notes that Thomas Gill, a gun and sword maker of Birmingham, patented a rifling machine in 1800 where:

> The rotation of the cutters is effected by a rack moving transversely to the bed which takes into a pinion on the end of the cutter bar. The extent of the traverse of this rack, and therefore of the revolution of the cutter and amount of twist given, depends on the steepness of an inclined plane which forms part of the mechanism.

Prosser also notes: 'Whether a machine on the principle of the patent was actually constructed at the time is somewhat doubtful'.

Writing on the Enfield in 1854, Jervis-White Jervis shows a rifling bench operated by two men. It clearly shows the use of a multi-point rifling file as the cutting tool, the cutting tool cutting both as it enters and exits the barrel.

He remarks:

I have endeavoured in this wood cut to give some idea of the labour required for rifling a barrel. But the enormous force required to push such a bit as is here represented, and the time taken up in so doing, naturally causes the looker on to wonder how it is that steam has not been introduced into this branch of the department. It would indeed be a matter of economy to do so, and, if I am not mistaken, the French have some such engine in use.

Wallis and Whitworth[68] describe USA Techniques in 1854:

All rifling is done by self acting machinery of simple construction. A bar of the required twist is placed behind or under the machine, and on this bar rests a roller connected

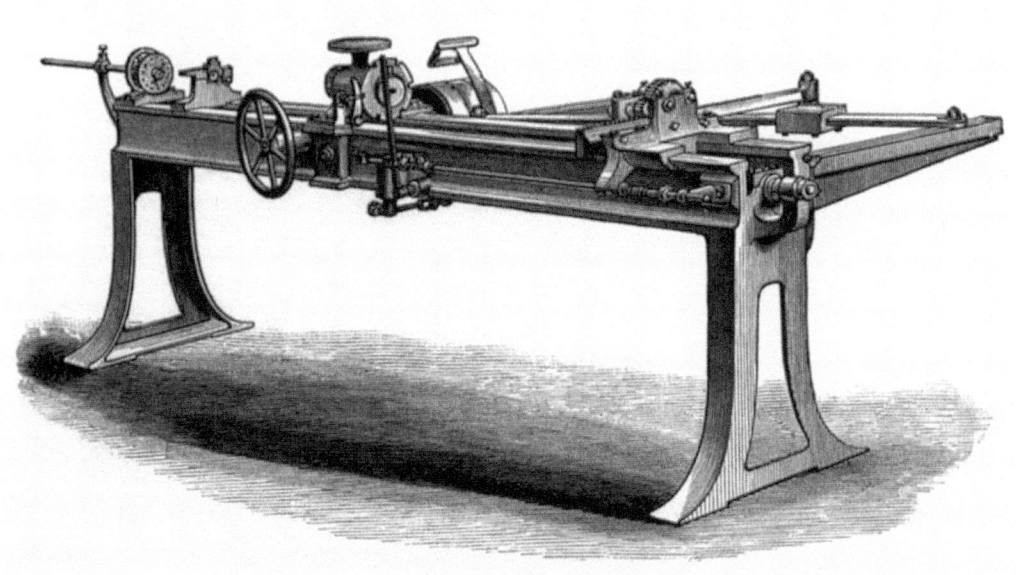

5.32 *Muir and Sons rifling machine. (Walsh 1884)*

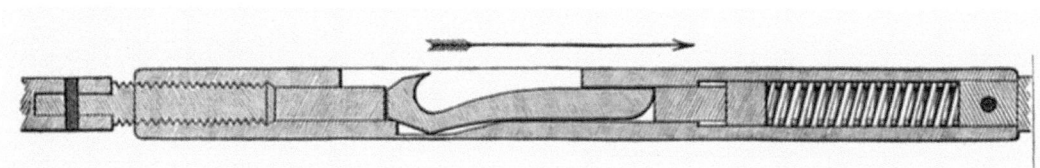

5.33 *A hook cutter. (By permission of the Muzzle Loaders Association of Great Britain (Walsh 1884))*

with a rack, as this rack is drawn along the bar it rises or falls according to the form of twist, and this rising or falling motion given to the rack acts upon the rifling bar, thus giving it the required twist; the rifling bar is worked by a crank, which does not appear to be so uniform as a screw would be for the same purpose. The barrel is turned round a division at every cut; this is accomplished by the motion of the rifling bar; which also puts a detent out and in to prevent the barrel from shifting.

While this machine includes a rack and pinion, it appears that the helix is still generated by following a bar rather than via any gear box.

The sine bar rifling machine – like the one supplied to Lovell and outlined by Gill – (**5.31**, designed and built by Greenwood & Batley, and **5.32**[69]) uses a reciprocating, hooked single point cutter – as shown in **5.33** – to cut a single rifling groove at a time. A geared mechanism coupled to a sine bar helps to drive the cutter and ensure the correct changing orientation of the rifling groove. The 'sine bar' uses two sides of a triangle to accurately set the helix angle. A form of dividing plate or protractor is used to ensure that grooves are in the correct position relative to each other. This method was extensively used at Enfield.[70]

The main machines used during the twentieth century[71] were as follows: sine bar hook cutting machines were manufactured in thousands in the First World War and need to take eighty passes for each groove; in the Second World War, more modern machines used a hydraulic drive and a large diameter thread as the generator of the twist. Depth of cut was still only one tenth of a thousandth of an inch. Alter-native approaches cut more than one rifling groove at once. Scrape cutter rifling uses a bar with curved and hardened steel scrapers set into it with one scraper for each groove. An increasing depth of cut is made on each pass of the multi-groove tool. Broach cutting cuts all the grooves to size on one single pass of the tool. The tool used has twenty to thirty stepped cutting edges that progressively generate the bore. Metal-forming processes are also used. 'Swage or hammer rifling' uses vertical or horizontal rotary swaging machines to hammer a tube radially over a fluted tungsten carbide former to create the rifling grooves.[72] Swaging gives a mirror finish bore and does not therefore require a finishing process. It is primarily a European approach. During the 1940s the cold forming process of button rifling seems to have really taken off in the USA. In this process, the 'button', a shaped end or projections on a rod, is pushed or pulled through a smaller diameter tube to form a rifled bore. Remington particularly worked to develop both the tungsten carbide button tooling and its lubrication during the forming process.

The Final Technical Steps

Other metal working machine tools with their genus in USA gun making in the mid- to late nineteenth century are: the turret lathe created by Stephen Fitch of Middlefield,

Connecticut, in 1845, initially applied in the manufacture of percussion locks for an army horse pistol;[73] its refinement, the automatic turret lathe that included as its key innovation a rotating drum with moveable strips on its surface to act as cams to drive the motion of the tools, developed in 1873 by Christopher Spencer, the inventor of the Spencer repeating rifle; and, indirectly, the truly universal milling machine with the classic 'knee and column' configuration created by Joseph Brown of Brown & Sharpe in 1862 (**5.34**). This was initially developed for the manufacture of fine twist drills such as those used to drill percussion nipples, the drill flutes were previously hand filed.[74] The machine was particularly novel in that it was not based upon the layout of the lathe. Brown & Sharpe also enabled the wide application of the familiar hand-held micrometer calliper, producing it in large volumes from the late 1860s. Accuracy in the micrometer is achieved by creating a precise screw thread and using *end* measurement against anvils – Whitworth's principle – rather than comparing the object being measured to the edge of a scale. Spencer also invented the Board Drop hammer, which uses a wooden board between rollers as a key feature of the drive for the hammer.[75] Again, the manufacture of firearms was a major factor driving innovation in metal forming.

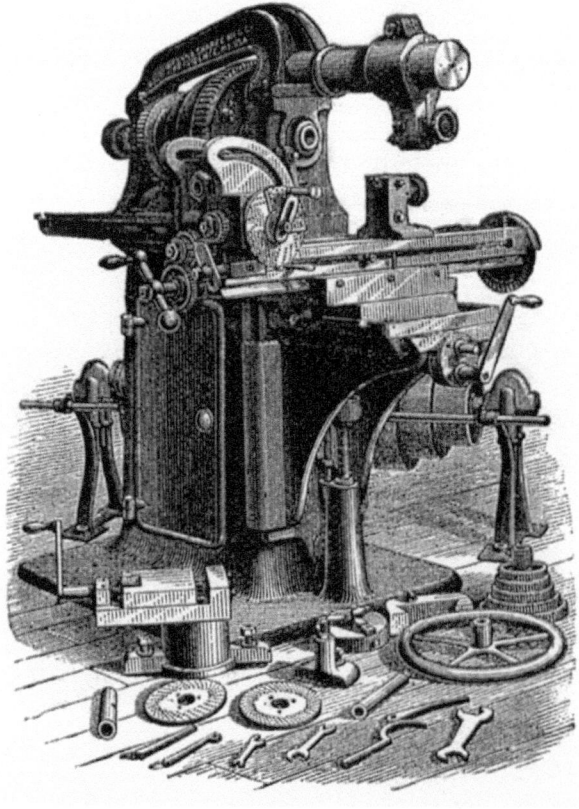

5.34 *Brown & Sharpe's universal milling machine from a trade advertisement of the end of the nineteenth century.*

In these machines and measuring systems we see a transition from special purpose to general purpose devices and the introduction of repetition processes. Schelsinger,[76] the pioneer of machine tool testing, tells us that in the first years of the twentieth century, grinding became the basis of interchangeability, with precision grinding tolerances of the order of +/-0.0002in in 1902. Grinding had been turned into a precision process by the invention of the bonded wheel in the F.H. Norton Co. in the USA. The bonded wheel was patented in 1887. Grinding was and is, especially important in the manufacturing of gauges[77] and tooling.

During 1910, the Institution of Mechanical Engineers held a large meeting in Birmingham, including visits to BSA and Webley and Scott. The notes of the visit to BSA[78] tell us the following:

> The limit-gauge system of inspection is used throughout the works. For coarser dimensions an allowance of +0.001in is generally permitted, but for all finer measurements +0.0005in and +0.00025in are the usual allowances, though in many cases this allowance is further reduced to +0.0001in.

We should note the use of limit – go, no go – gauges rather than receiving – go – gauges. BSA were also still using Ryder forging machines. Forging tolerances for gun making were 0.002in in the USA, primarily because inspectors were applying the tolerances they understood from metal cutting as above.[79]

The gun making industry was progressively generating many of the machines that were exploited within the volume manufacture of the consumer goods that came with the closing of the nineteenth century, particularly the sewing machine and the bicycle[80]. There was still one significant step to be made. It was made by Carl Johansson in the Eskilstuna Rifle Factory in Sweden.[81] Johansson was chief armourer and had the task of tooling up in the 1880s to make the Mauser rifle.[82] Johansson visited the Mauser works at Obendorf and saw the thousands of gauges needed to make the rifle. Examining a rifle of this period shows that they are covered with inspector's stamped view marks, each one corresponding to a gauging or other viewing step. While he was there, he was inspired by the multiple length gauges of the Polhem Stick from the 1700s – **5.35** shows how a single gauge could be used to measure a number of different lengths from steps cut into the gauge. Each step is located at one end of the dimension to be measured and the thumb used at the other in order to feel any differences between the gauge and the work piece.

> Christopher Polham, a Swede, devised many techniques for the application of machinery to the quantity production of metal and metal products, but could not successfully implement his conceptions with the power sources and clumsy wooden machinery of the first half of the eighteenth century.[83]

5.35 *Polhem sticks. (Reproduced by permission of Wayne Moore from the Foundations of Mechanical Accuracy (1970))*

His gauges were first made of wood and then of metal. Johansson conceived the alternative, less expensive, idea of a set of 102 gauge blocks that, with a length comparator, would allow the making of 20,000 different measurements. A comparator allows a piece of work to be compared with a gauge, the most well-known is a dial test indicator or 'clock'. Clearly the accuracy of the comparator must match that of the gauges for the results to be meaningful. Johansson's blocks were very precisely manufactured (Johansson's measuring accuracy was 0.001mm, 40 millionths of a inch) and finished such that they would *wring* together, being held only by adhesion forces. These gauges where still considered novel in 1912.[84] Johansson was also one of the major pioneers of measurement standardisation within the workshop and to national and international standards. Measuring to international standards that ensure a shared definition of lengths allows complete interchangeability.

The gun-making industry, with its demand during the First World War for significant numbers of accurately machined jigs, led to the creation of the jig borer. Jigs are used to accurately locate drills during the drilling process, so that each component is produced with the holes in the correct position. A hardened bush guides the drill. Jigs are usually distinguished from fixtures: jigs are used to guide drills, fixtures are used to hold components for cutting processes. There were two processes to set these drill hole positions, the use of a master plate that copied the positions but did not ensure that the dimensions were correct, or the use of a 'button' and 'wiggler'. A tool maker's 'button' is a hardened roller assembly screwed into a tapped hole, close to the correct position of the hole. The button is adjusted so that it is in the correct position on the work and clamped tight. A wiggler is essentially a lever that magnifies displacement at one of its ends. The wiggler is held on the boring head of the machine tool and set to touch the button. The boring head is rotated and the end of the wiggler watched to see if it moves. If it moves, the relative position of the button and wiggler are not correct. The toolmaker then moves the position of the work to minimise the amount of wiggle, removes the button and carries out the boring process. This is a very time consuming process! A machine was therefore necessary to allow the rapid accurate relative positioning of holes and then to bore these holes with the work in the same location. The machine therefore combines the features of an accurate measuring machine or instrument with that of a machine tool. This innovation 'was late in coming'[85] and was invented by a Swede, B.M.W. Hanson at Pratt & Whitney in 1917. Pratt & Whitney were in the process of equipping many armouries and had a need to manufacture large quantities of jigs, fixtures and gauges. The jig borer has both a highly accurate boring spindle and mechanisms, micrometers using end measurement, so that the slide ways of the machine can be accurately and quickly set in particular positions.

In this chapter we have seen how many of the core techniques of conventional mechanical manufacturing have their roots in gun making and how these techniques impacted on one of the main centres of handicraft-based manufacturing. These new techniques, while damaging one part of the manufacturing base of Birmingham, recreated it both by the impact of the new BSA factory on gun making itself and, on increasing the knowledge of the techniques understood by the manufacturers of Birmingham as these diffused – relatively slowly when compared to the USA[86] – from BSA to other industry sectors, most significantly to the manufacture of cycles. It is also argued[87] that Birmingham was fortunate to not have a large railway works as the 'strong individuality of the chief engineers' constrained the rate of change to the new technologies.

The reader more familiar with the evolution of firearms will have noted while reading through the book the relationship between the innovators in manufacturing techniques for weapons and ammunition and the innovators in the design and successful exploitation of breech loading. These include Hall at Harpers Ferry and his breech loader; Robins and Lawrence and Sharps; Howe, the Providence Tool Works and the Peabody breech loader and Spencer with the automatic lathe and his revolutionary breech loader.

6

THE BIRMINGHAM TRADE FROM 1880

This chapter reviews the progress of the trade from the close of the nineteenth century to the present day with a special focus on the Depression of the 1920s and 1930s. It identifies the role of machinery in the twentieth-century sporting gun trade and shows photographs of the St Mary's area in the 1950s and the trade today.

Machinery and the Manufacture of Sporting Guns

While the use of machinery in military gun making was ubiquitous by the 1870s, it was taken up more slowly by the sporting gun trade because of its strong craft tradition. Progress is succinctly described by Richard Akehurst[1]:

> As gun actions and mechanisms became more complicated that advantages of machines to do some of the basic work was increasingly appreciated, but all the finishing and fitting was still carried out by skilled specialist craftsmen. When the moderately priced ready made gun began to be produced in large numbers, especially the cheaper box locks, standardisation and the use of machinery were of such benefit in reducing costs that enterprising gun makers who specialised in this class of work were quick to see its advantages. There was never the less, even in these guns, still a fair amount of hand fitting and finishing.

6.1 (Above and Right) *Victorian treadle lathes.*

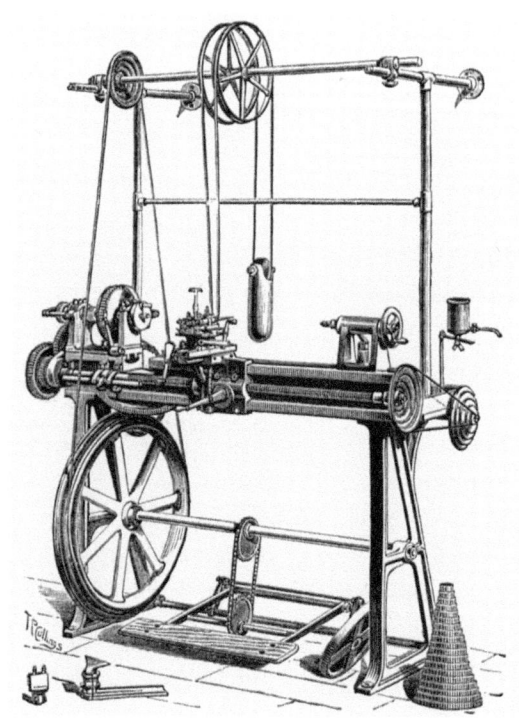

Walsingham and Payne-Gallawey wrote in 1887 that actioning the gun 'is one of the most expensive parts of the building of a gun – and in a hammerless one *the* most expensive'. One such enterprising gun maker was W.W. Greener, who in the 1910 edition of his book, *The Gun and Its Development*,[2] details the advantage of the machine:

> In the production of sporting guns machinery now has an important part... Much of the labour which was formerly done by hammer, file and foot lathe is now more readily and cheaply produced by the judicious use of machinery. In the hammerless gun, for example, much work that can be done indifferently by hand-drills, 'routers' and special tools is cleanly and squarely cut by slotting and profiling machines... The work that can be profitably accomplished by machining includes much of the first shaping of breech-actions and barrels, but it is only of late years that steam machinery has been considered necessary for the manufacture of sporting breech actions... The milling machine, which does most of the work to the breech actions resembles a lathe head on a short bed [i.e. a manufacturing miller], in front of which is a slide rest capable of moving vertically and horizontally.

6.1 shows two Victorian foot or treadle lathes. This application of machines made the subsequent fitting processes much easier and reduced costs, thereby making the guns more accessible, machinery being accepted by gun makers with a view of economic reality, whereas high value specialist gun makers tended to emphasise the value of hand working and were vertically integrated to include all the skills required, rather than using the network of sub-contractors and outworkers.[3]

Leslie Taylor,[4] Managing Director of Westley Richards at the turn of the century, echoes the benefits of these new approaches and their relationship with older practices, 'Machinery has stopped waste of muscle and patience... saving grinding labour... Gun making still requires the skilful handiwork of trained men, which no machine can be made to imitate or replace.' Westley Richards had invented the Anson and Deeley hammerless action in 1873 dramatically simplifying action manufacturing by reducing the number of parts by fifteen.

Materials innovation also progressed. Birmingham twist and Damascus barrels used in the sporting gun trades had visual faults called 'greys' – pieces of scale embedded during the welding process. Belgian barrels were soft and could bulge at the choke[5] on proof, and were used in three quarters of the Birmingham production. J.H. Walsh, the sporting gun writer who used the pseudonym Stonehenge, describes[6] Mr Smith Casson, Manager of the Earl of Dudley's Round Oak Ironworks[7] and Thomas Webley, visiting the Belgian barrel makers to learn why they had less visual faults and why the lower quality Belgian barrels were so soft. The visitors found that the Belgians worked their material much less than those in Birmingham and that the mechanical working was done much later in the process, a mandrel being required to transport the soft tubes. They also used a smaller (and cleaner) forge fire. Casson and Webley improved their own processes for barrel making after the visits but did not share the steps that they took with Stonehenge. Progressively, barrels were made with steel, first Whitworth's fluid compressed steel and then Siemen's process steel. This improved material allowed the forging out of the 'chopper lump'[8] at the breech end of the barrel. This removed the step of soldering or brazing the lump to two barrels, leaving just the complex task of joining the two barrels.

At the turn of the century, the debate on the use of machines in the making of 'reach me down' or ready made guns – 'excellent *as far as they go*' (Walsingham and Payne-Gallawey, their italics, 1887) – seems to have reached its peak. It was recognised that most guns would be, to an extent, machine-made but that the best gun – because of its fitting to the individual[9] and the demand of the market for fine finishes – would require more purpose built and fitted parts with high finishes. It has been suggested[10] that the trade might have accepted machinery more readily if it had been based 'close to the machine-producing centres of Manchester, Leeds or Glasgow'. However we can see, from the machines being produced by the local machine builder Archdale, that there was a local supplier able to meet a local demand. The growth of other machine-based industries – such as the automotive industry in the West Midlands – also supported the application of machines to the industry.

The Contraction of the Birmingham Trade

The introduction of machines was clearly a significant step in the decline of the Birmingham craft gun-making trade, although those companies that adopted them and applied them to the production of military magazine rifles and machine guns prospered. This, and tough competition from European manufacturing centres such as Liège and the new North American centres, continued to progressively erode the scale of the manual and sporting gun trade, combined with economic depressions including that of the decade until 1886. Trade barriers, such as the protectionist McKinley Tariff imposed in North America, made it difficult to export low cost weapons, although high quality arms continued to maintain their markets:[11] 'Many a tale is still told in the trade of gun men working all night by candlelight to prepare guns for the last boats to sail before the imposition of the tariff'.[12] The Greener's as Artifex and Optifex, recorded the gun trade exports to America, which fell from $1,169,000 in 1882 to $349,000 in 1890, to less than $20,000 in 1905. The number of people employed in the trade fell from 5,500 in 1881 to 4,100 in 1911.[13] The trade continued to contract after the First World War, and was hurt by the Depression. This continued decline is decried by C.E. Greener in his introduction to the Birmingham Proof House History written in 1949.[14] Continuing legislation to constrain the ownership of weapons within the UK has also had an impact on the trade, as did air-raid damage in the Second World War,[15] the adjacent Snow Hill station being bombed in November 1940 and April 1941.[16]

THE BIRMINGHAM TRADE FROM 1880

BSA and Gun Making in the Twentieth Century

BSA were still using many of the early machines to make weapons during the First World War, and had also made many weapons during the Boer War. At the close of hostilities, demand reduced and BSA turned to making small quantities of sporting rifles and machine-made shotguns to attempt to use some of its excess capacity. It progressively turned from a gun-making business to apply its interchangeable manufacturing skills to bicycles, motorcycles (some of the gauges shown in **5.4** are for motorcycle components) and cars, including the acquisition of Daimler. It became a group with three product strands: guns, cycles and machine tools, including special purpose machines. With the coming of the Second World War, arms and transport manufacture once again became critical and BSA controlled sixty-seven plants in all (its own and shadow factories), 28,000 people and 25,000 machine tools, and produced more than half of the small arms supplied to Britain's armed forces. The group continued to grow after the Second World War, including the significant step of becoming the largest machine tool organisation in Britain by the mid-1960s and accounting for 80 per cent of British motorcycle exports by 1969. It also 'maintained a steady flow of air rifles, hunting rifles and sporting guns to most of the world'. Unfortunately, the group performance began to deteriorate at the end of the 1960s, with significant consequences on motorcycle and machine tool businesses. In 1975, after many struggles, the motorcycle business was closed. Some gun making continues but at a much reduced level. Gun making, especially military gun making, is clearly and perhaps fortunately a difficult business to sustain.

The Carr Brothers and 10–11 St Mary's Row

A typical story of the contraction of a small business and its workshops in the sporting gun trade in the twentieth century can be made by combining material from a number of sources[17] on the business of the Carr Brothers, who operated from 10–11 St Mary's Row, to relate the following: James Carr (1885–1895) & Sons (1895–1900) operated from 10–11 St Mary's Row, Birmingham. Joseph Westwood Carr was a Guardian of the Proof House.[18] The Proof House records show that Joseph was elected as a Guardian in May 1938 and served until to 1959. Walter John Carr and James Stephen Carr were also recorded as master gun maker and gun maker respectively from 1912 to 1937 in the Proof House Register of the Birmingham gun makers. The last Carr in the register is Mrs C.S. Carr, who is recorded as a Director of William Ford from 1962.

Their workshops[19] are shown before their demolition in **6.2**, and in plan in **6.3**.

> Numbers 10–11 St Mary's Row comprise a three-storey brick building… fronting directly onto the pavement. The front elevation is typical of eigh-

teenth-century architecture. The name plates of the firms in occupation are the sole external indication that these are industrial premises. The central doorway leads into a passage providing access to the workshops in what was originally the back garden.

In the old garden there are two 'courts' of workshops separated by a three-storey building, again 'pierced only by a narrow passage'. The upper workshops are reached by external 'outside staircases' as these 'had the advantage of not taking up any floor space in the workshops which were in any case very small'. Space is at a premium: 'It was necessary to build as high as was possible without obscuring other workshops from the daylight'. Turning to the workshops themselves:

> Particularly significant is the size of the windows[20] which run almost the entire length of the workshops – some indication of the importance of a room with plenty of light to the trades metal workers and engravers... The back garden workshops have not always been in the gun trade... in 1845... Joseph T. Horton, ivory, bone and hardware turners were in occupation and in 1856 it was George Tonks manufacturer of chandeliers and gas fittings.

Correspondence with Bill Roper, who served part of his apprenticeship seconded from London to the Workshops at Nos 10–11, shows that not all in the buildings were gun makers in the 1950s: 'A firm of copper cylinder makers were in occupation in St Mary's Row who had to store the finished product on the pavement outside because of lack of space'.

An analysis[21] of two of the Carr's cash books summarises the progress of the gun trade during the 1920s and 1930s. The books run from July 1923 to December 1928 and from May 1930 until December 1935. From a business containing three Carrs, a wages bill of £60 and giving work to thirty men in the 1920s, to a business in the 1930s with one Carr, a wages bill of £25 and only giving work to four or five other men. **6.4** shows the Carr's trade appointment card from the 1920s.

In 1948, two gun makers, three gun action makers, one barrel browner and rifler and one engraver and polisher were in the workshops at 10–11 St Mary's Row.[22] Barrel makers William Ford amalgamated with James Carr & Sons in 1953, after the death of the last William Ford in 1946.[23] The William Ford business worked from 15 St Mary's Row[24] between 1889–1948, moving to the adjacent Price Street in 1948 for a short period before returning to St Mary's Row at 10–11 on the amalgamation. They became a limited company in 1954. In 1965 they moved to Potters Hill in nearby Aston and then to 352 Moseley Road on the other side of the city centre, close to Camp Hill. They disappeared sometime in the early 1990s. A comment about the tenants of the workshops at 10–11 just before their demolition in 1965 is: 'A short time ago there were still three firms describing themselves as gun makers, one repairer and one gun engraver...'[25]

THE BIRMINGHAM TRADE FROM 1880

6.2 (Right and Below) *The exteriors of 10–11 St Mary's Row before their demolition.*

THE BIRMINGHAM GUN TRADE

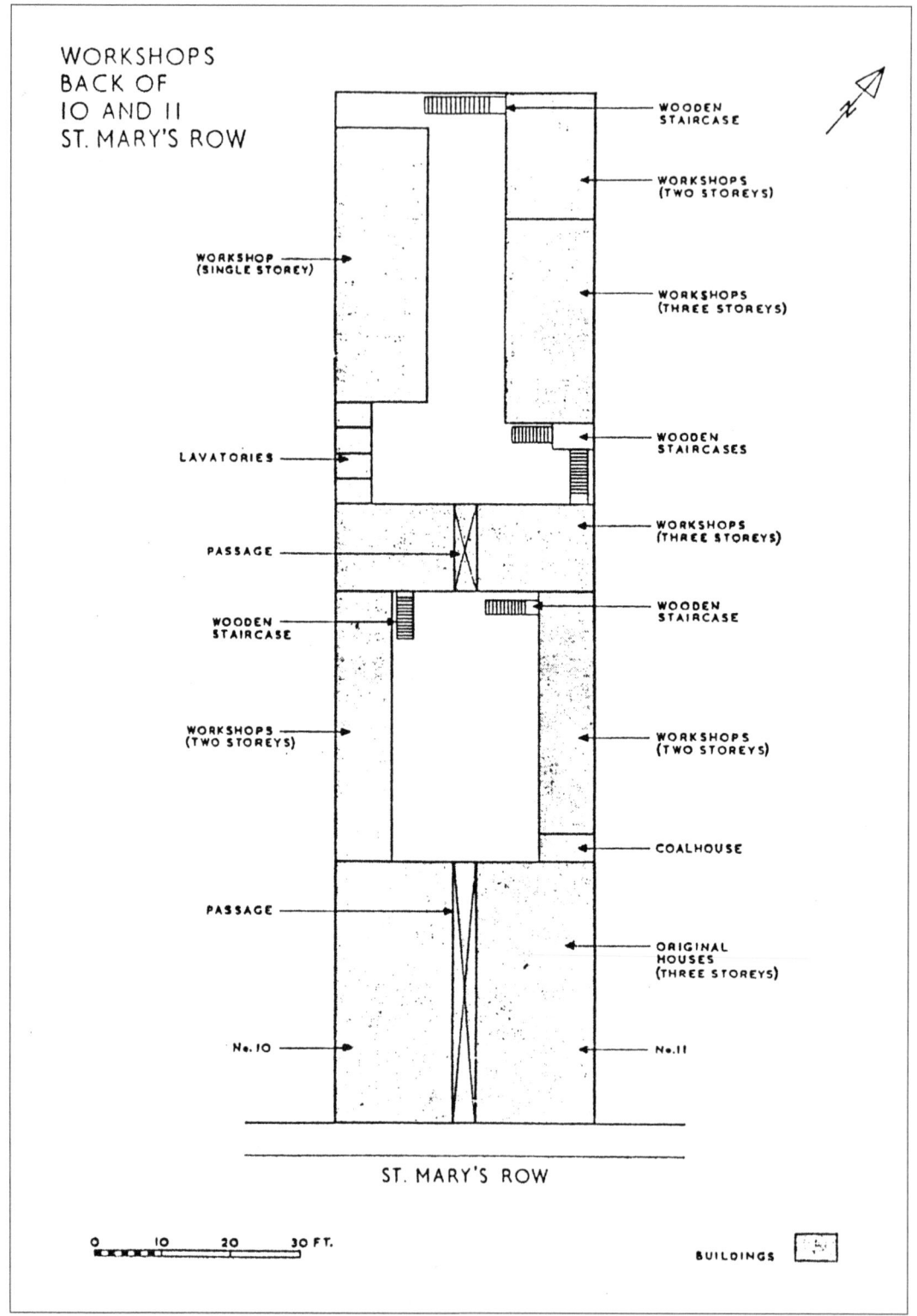

6.3 *Dr Smith's plan of 10–11 St Mary's Row.*

THE BIRMINGHAM TRADE FROM 1880

Bill Roper describes some of the tenants in the 1950s:

> The two-storey workshops in the first courtyard were occupied on the ground floor by William Woodward, a polisher, colour hardener and charcoal bluer who prided himself on the lack of distortion in the finished product. Directly opposite the main door to these workshops and Woodward's shop were the stairs, about 18in wide and still with the brackets fixed to the wall that had held the oil lamps to light the way. Under the stairs was a fixed wooden seat and the fixings that had held a crank, turned by an apprentice boy to power the fine borer by flat belt drive in the workshop above, this workshop was tenanted by Thomas Yates a barrel borer, dent raiser and general barrel repairs. The seat was made redundant upon the installation of electricity.

Dunham wrote of the contracting trade in 1955:

> The old gun trade is nowadays very restricted in the matter of supplies, and stampings and forgings for guns are becoming very scarce indeed. Even more critical is the supply of tubes for shot gun barrels. With the exception of imported tubes there is virtually only one source for these.[26]

The final step in the contraction of the trade was made in the development of the Smallbrook Ringway in the 1960s, leading to the demolition of many of the traditional workplaces. A centre spread in the *Sunday Mercury* of 24 February 1963 announced the 'Death of the gun quarter'.

Boothroyd gives us an atmospheric description of one of the workshops in the 1960s in *The Handgun*:[27]

```
                        MEMORANDUM.
                        ──────────
        From  JAMES CARR & SONS,
                        St. Mary's Row, BIRMINGHAM.
        ════════════════════════════════════════════
                                          ..................  ........192

        Our Mr....................hopes to have the pleasure of calling upon

        you on or about ..........................................................

        ...........................................................................

        when your favours will oblige.
```

6.4 *The Carr's appointment card from the 1920s.*

6.5 The Gunmakers Arms today.

I became quite friendly with Mr Morris, the engraver and his shop, typical of the small Birmingham workshop, had remained unchanged for over half a century. It was approached through the courtyard and up two flights of outside wooden stairs which were a danger to life and limb, particularly after dark. Lit by gas and heated by a cast iron coke stove, the workshops were lined by benches on which lay a profusion of tools and of cardboard and tin boxes, all in a state of indescribable confusion. The proprietor had his own bench against a window with a north light and, after dark, work would be carried on by the light of an incandescent gas mantle, the glare softened by a mask of tissue paper... My visits were made in the twilight of the British gun trade... Many of the specialist workers in the trade had left... but in the evenings could earn extra money at their old craft and, from five o'clock onwards, men would pass into the workshop from the twentieth century and each would go to his bench... At nine o'clock the files and gravers would be laid on the bench, and coats and hats would be put on in time for a glass of beer at the local before closing time.

6.5 shows the Gunmakers Arms, one of the locals; another was The Bull at the corner of Price Street.

Bill Roper confirms that Harry Morris was at the top of the 'Three Storey Workshops' for the best light for his engraving, 'but whenever I saw him working he was using the shaded gas mantle which he said was not quite as good as the tallow dip for throwing a shadow on his work'. Bill also tells us that Harry was old when he knew him and 'unable to carry the ashes from his stove down the outside staircase and so stacked them around the walls of the anteroom... to be disposed of by others when the weight threatened the structure'.

THE BIRMINGHAM TRADE FROM 1880

Photographs of the St Mary's Workshops

During the closing years of the 1950s and early 1960s, the Birmingham Museums and Art Gallery took a number of photographs of the Birmingham gun making workshops. The staff of Birmingham Museums and Art Gallery, particularly Dr Jim Andrew, have been most helpful in tracing and providing access to these images. These are reproduced here with their permission. A number of these images have already been used in Chapter 2, showing bench work, **2.3**, and barrel boring, **2.20** and **2.21**.

The photographs can be viewed in a number of groups. **6.6–6.9** show the small workshop of Robert Kyte. **6.6** shows Kyte at his bench, working on a barrel in the light of his window. Through the window we can see an external staircase. Around the walls we can see hand tools and gauges and a cast off stick. The photograph above the clock is of the workmen at Greener's[28] factory. **6.7** shows the bench from another angle. **6.8** shows the other end of the workshop in the evening: the gas mantle is on, Kyte is pumping the bellows of his forge and behind him is his treadle lathe. **6.9** shows us a better photograph of his lathe. A chimney from the stove winds across the workshop – perhaps the factory inspectors had been in to all the workshops and required the fitting of these chimneys – recur in the images.

6.10 and **6.11** show the larger workshop of J.K. Pople displaying the use of machines, including a number of treadle and other lathes and a large pillar drill set up for a special task. All the machines are driven by line shafting. **6.10** shows the benches, one supporting a fly press. In **6.12** we can see a veteran milling machine machining the lumps of a pair of barrels held in a machining fixture.

6.6 *Robert Kyte at the bench in his workshop, 1959. (By permission of the Birmingham Museums and Art Gallery)*

THE BIRMINGHAM GUN TRADE

THE BIRMINGHAM TRADE FROM 1880

6.9 *The treadle lathe in Kyte's workshop, 1959. (By permission of the Birmingham Museums and Art Gallery)*

Opposite Page

6.7 (Above) *Another view of Kyte's bench, 1959. (By permission of the Birmingham Museums and Art Gallery)*

6.8 (Below) *Kyte operating the bellows of his forge, 1959. (By permission of the Birmingham Museums and Art Gallery)*

6.10 (Above) *A sporting gun workshop showing the use of machines, including treadle lathes, in the 1950s. (By permission of the Birmingham Museums and Art Gallery)*

THE BIRMINGHAM TRADE FROM 1880

6.12 *Machining lumps in the 1950s. (By permission of the Birmingham Museums and Art Gallery)*

Opposite Page

6.11 (Below) *The benches at the side of the workshop in the 1950s. (By permission of the Birmingham Museums and Art Gallery)*

6.13 *The slotter in the 1950s. (By permission of the Birmingham Museums and Art Gallery)*

THE BIRMINGHAM TRADE FROM 1880

6.14 *A hearth in the 1950s. (By permission of the Birmingham Museums and Art Gallery)*

6.14 and **6.15** show us an open forge with some 'Vargas' calendar pin-ups and browning or bluing in an enclosed box on top of a coke fire respectively. Perhaps it was not always necessary to stoke up the stove in **6.15** and we can certainly understand why the cat made itself comfortable here! **6.16** shows us finishing on the 'mop' with a pot of polishing compound and stick to apply it to the wheel and a boy working at the hearth. **6.17** is a lone man at his bench, perhaps an engraver.

6.18 shows us the rear of two buildings in Price Street before the demolition of the lower height workshops.[29] Within these workshops was the business of Malcolm Cruxton (now relocated in Price Street and still operating). **6.19** – **6.22** show the making of sporting guns in the workshop of Malcolm Cruxton in 1981, before the demolition of the building. **6.22** shows Malcolm with a 'horse', a wooden steady, on each side of his vice. **6.23** is the workshop of a barrel solderer, formerly across the road from Cruxton.

6.15 *Barrel browning in an enclosed box in the 1950s. (By permission of the Birmingham Museums and Art Gallery)*

6.16 *Finishing in the 1950s. (By permission of the Birmingham Museums and Art Gallery)*

6.17 *Bench work in the 1950s. (By permission of the Birmingham Museums and Art Gallery)*

6.18 *Price Street before the demolition of the workshops.*

THE BIRMINGHAM GUN TRADE

6.19 *Making sporting guns in the workshop of Malcolm Cruxton in Price Street, 1981.*

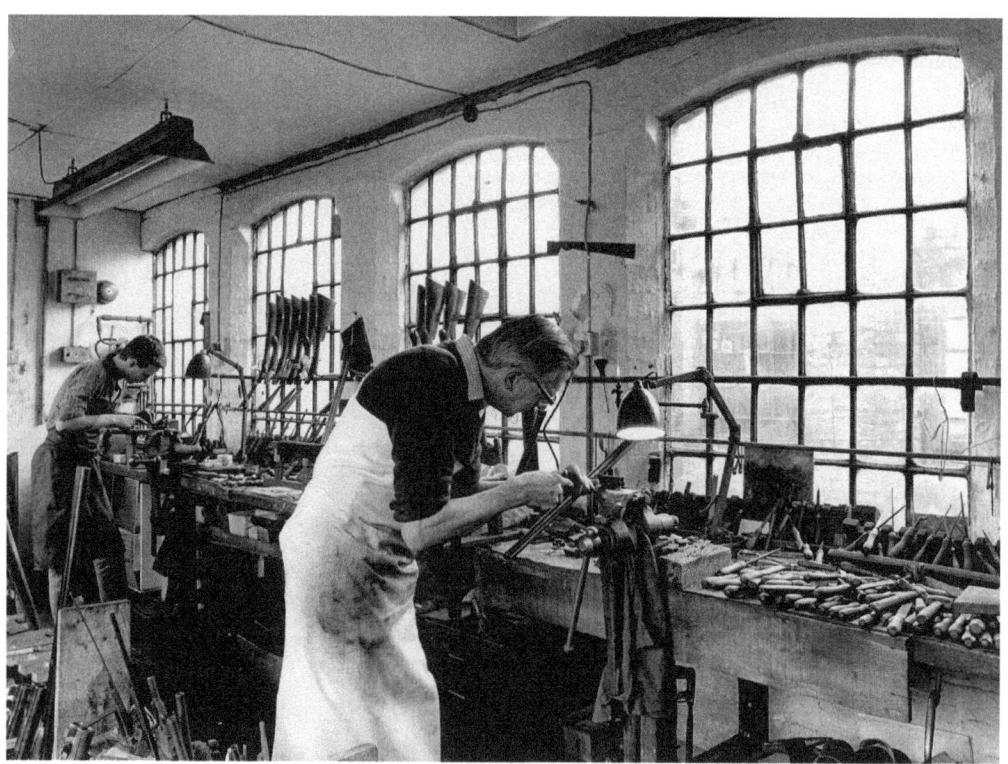

6.20 *Bench work in Cruxton's workshop, 1981.*

THE BIRMINGHAM TRADE FROM 1880

6.21 *Another view of bench work in Cruxton's workshop, 1981.*

6.22 *Cruxton finishes the 'teardrop' on the stock near the tail of the lock plate.*

THE BIRMINGHAM GUN TRADE

6.23 *Barrel soldering in 1981.*

The Trade Today

A little of the Birmingham trade still survives in its St Mary's location and scattered around the city, but much of it has disappeared or has relocated to other locations outside Birmingham. The trade has tended to migrate from gun making to the provision of gun repair and the sales of sporting guns and clothing. **6.24** shows Price Street today – the Cruxton business still survives on the right-hand side of the street. The building in the foreground is 'New Buildings', the higher of the two buildings in **6.18**. This was refurbished in 1980 and was previously the Partridge Works of Lightwood & Sons Ltd – photographs taken before the refurbishment clearly show this. It is thought to have been built by Bentley & Playfair during the American Civil War.[30] It still contains a number of gun makers, (see the board in **6.25**). The shooting of reproduction muzzle-loading guns also has given some impetus to the trade. Parker Hale manufactured reproduction 1853 Enfields until they were wound up in 2001. This work has now slowly migrated to Italian gun-making firms.

THE BIRMINGHAM TRADE FROM 1880

6.24 *Price Street today. Cruxton is still trading in a new location in Price Street.*

6.25 *New Buildings, Price Street.*

6.26 *Haydn Hill by the bench of his Stirchley workshop. (By permission of Haydn Hill)*

6.27 *Machines in the workshop. (By permission of Haydn Hill)*

THE BIRMINGHAM TRADE FROM 1880

6.28 *Another view of the rear of the workshop. (By permission of Haydn Hill)*

Individuals are also trying to grow and retain – with some success – the craft skills of high quality sporting gun making. One of these is Graham Greener (see his book[31]) celebrating the history and the future of a great gun making enterprise. **6.26**, **6.27** and **6.28** show the present-day bench and workshop of Haydn Hill, action maker, in Stirchley. **6.29** shows a milling machine set up for machining barrel lumps in a fixture similar to that shown in **6.12**. **6.30** shows a lathe set up for chambering a barrel and **6.31** a simple vertical miller. The Hill family business moved out of St Mary's as it contracted and they recreated their workshop at a new site. The Hill's workshop shows the role of machines in the manufacture of sporting guns in the traditional twentieth-century trade. Haydn still creates the fence of the action using a cold chisel.

The innovative business of Westley Richards is, as Colonel Hawker said in 1838, 'quite the star of Birmingham' and the home of the Anson & Deeley action in the 1870s. It still remains as a best gun maker[32] (see its trade advertisement of the 1880s in **6.32**) and continues to emphasise the value of craft skills complemented by precision machinery and measurement. This business was sustained through a difficult period by the repatriation and sale of classic weapons from India. The India pattern Brown Bess in **1.1** and **1.2** is one of these! **6.33** shows the engraving of a best gun today, and **6.34**, one of the characteristic highly finished Westley Richards double guns with patented detachable locks.

THE BIRMINGHAM TRADE FROM 1880

6.30 (Above) *A lathe set up to chamber barrels. (By permission of Haydn Hill)*

6.31 (Right) *A simple vertical milling machine. (By permission of Haydn Hill)*

Opposite Page

6.29 *A milling machine set up with a fixture to machine lumps on sporting gun barrels. (By permission of Haydn Hill)*

THE BIRMINGHAM GUN TRADE

6.32 (Above) *A Westley Richards advertisement from the 1880s.*

6.33 (Left) *Engraving a best gun at Westley Richards. (By permission of Simon Clode)*

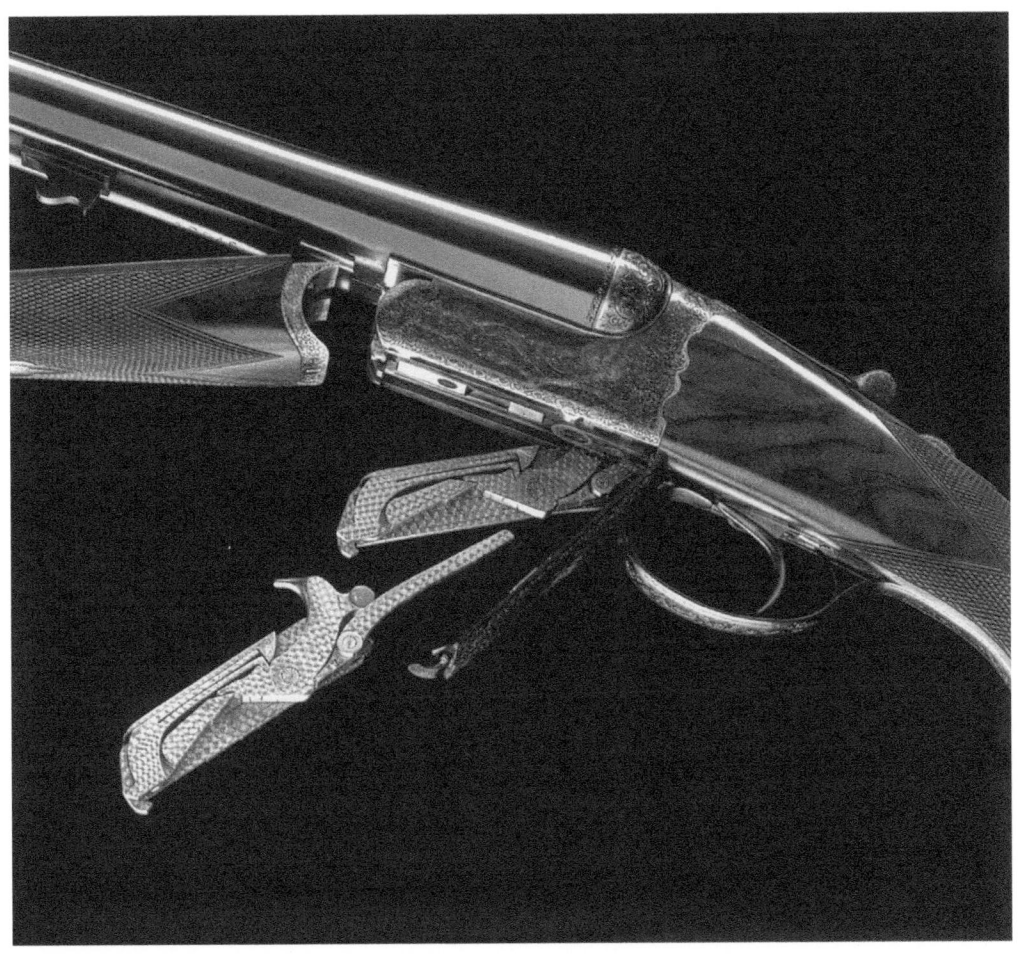

6.34 *A Westley Richards with detachable locks. (By permission of Simon Clode)*

CONCLUSIONS

As Roe[1] observed in the development of interchangeable manufacture: 'Gun makers were by no means the only ones to have a part in this development, but they were its originators, they determined its methods, and developed most of the machines typical of the process'. The machines invented by the gun makers were very quickly applied to the mass manufacture of the bicycle and sewing machine and subsequently the motor car; they remain the mainstay of the manufacture of conventional mechanical engineering products.

The Birmingham approach to gun making in the 1700s – that of the extreme division of labour combined with the craft skill of the artisan – should be recognised as a very significant step to enable mass manufacturing without machines. This pre-industrial or pre-factory approach is sometimes called 'proto-industrialisation' by authors in the political and social sciences[2] as they relate it to subsequent industrialisation steps that emphasise the role of capital, i.e. machinery. In Birmingham we can see the progressive replacement of the manual, craft-based domestically organised military gun making trade (excluding barrel making) by a factory and machine-based (capital-based) trade in a different location in the city – 'the transition from the workmanship of chance to the workmanship of precision'. In the first instance, the machines used are imported, these are then rapidly replaced by the use of indigenous machines in the same location – the leadership in machine tool making moving from the UK to the USA, the UK subsequently building incrementally on the USA innovations. The manual trade, while it still survives, has focussed on high value sporting guns and has been progressively eroded over the years. As the experts on the Birmingham trade, De Witt Bailey and Douglas Nie[3] have said:

> The trade... was able to produce surprising quantities of arms in a pre-technological, non machine oriented complex through the device of extreme specialisation of (craft-based) labour... the need for skilled labour was gradually reduced by firstly the invention of various machines for barrel making, secondly by the general acceptance of the percussion system, and lastly by the general introduction of machinery for making interchangeable parts.

Sustaining the capital requirements of such a business is clearly problematic as the experience of Colt and BSA shows – unquestioning application of interchangeable manufacturing techniques without fitting can be economically not sustainable.[4]

I am sure that the debate on machine-made versus hand-made, or at least hand finished and fitted sporting guns, continues today. As Dunham said in 1955: 'Birmingham and its gun makers have never taken to the idea of making shotguns by machinery. The complex nature of the mechanism and the narrow margin of safety left in a lightweight sporting gun call for the utmost accuracy of workmanship'. James Howe, at the close of his magnificent book *The Modern Gunsmith; A Guide for the Amateur and Professional Gunsmith in the Design and Construction of Firearms, with Practical Suggestions for All Who Like Guns*[5] has a chapter 'Appraisal of Craftsmanship' essentially comparing machine-made American guns with the hand finished and fitted British guns as follows:

> When we chose the very best we must go to England and select a British gun for we have not developed that high quality of workmanship which is so essential in fine arms... such close workmanship... is obviously impossible in cheaper grades of machine-made production.

Clearly machines *can* now be made to perform with such precision. However, we do not apply these machine construction techniques, as used for example in silicon fabrication or optical device manufacture, to feature sizes of micron dimensions to the best gun. The cost would be impracticable and the handicraft product (or at least the handicraft finished and fitted product) still has real value in the marketplace. We should recognise that in both the UK and the USA the value of the hand-made gun is still recognised by the sportsman and enthusiast. As Westley Richards say in their year 2000 catalogue[6] on the relationship between the machine and the craftsman being both a gun maker and maker of highly specialised precision press tools:

> Westley Richards is widely recognised for the quality of its engineering products. As a result the company has invested in modern computer numerically controlled milling machines, wire and spark eroders, which are also used to produce the blank compo-

nents of a gun's mechanism. Although these parts are produced to a degree of precision and finish that would satisfy many companies – and even some gun makers – Westley Richards regard them as merely a starting point in the process, requiring many hours of hand fitting and finishing before they can become an integral part of one of their guns.

In 2008 Wesley Richards relocated to Pritchett Street, at the centre of the Trade, recreating their traditional gunmaking workshops in combination with a retail outlet and a purpose built precision tool making business.

Recent conversations with Richard Purdey, former Chairman of Purdey, allow us to understand more fully what happened following the disappearance of the Birmingham gun component suppliers. Initially precision engineer entrepreneurs saw the opportunity to manufacture and supply CNC machined parts to the craft gun trade. Unfortunately this business was difficult to sustain and subsequently the best gun makers including Purdey and Holland and Holland and, as we have seen, Westley Richards created their own small internal CNC machine shops to supply components to their craft trained gunmakers.

The sporting shotgun that is specifically fitted to its owner to allow rapid mounting and aiming without the use of sights requires the use of stocks that are individually fitted to the individual. This in turn requires the tuning of the metal parts to fit the stock. This is not necessary in the robust military weapon which also has adjustable sights. The technological journey first gave interchangeability of form (by the special purpose machine and by the receiving gauge), fit (by accurate measurement) and is now working to give consistent function and finish with a price that the market will bear. This challenge remains in gun making and in the manufacturing of other artefacts.

One of the key innovations that led to interchangeable manufacture was the use of gauging, particularly sets of receiving gauges in the United States, the technology of gauging taking some time to find wide application in the UK. This highlights to us both the importance of the technical step and the necessity to rapidly diffuse the technical step within the industry and related industries to maintain the competitiveness of those industries.

If we visit Birmingham now we can see that much of the gun-making industry is gone, as have many other Birmingham industries over the years. Other examples of such dramatic rises and falls in Birmingham are those in bicycle and motorcycle making and more recently the contraction of the once large West Midlands-based automotive industry and the very significant machine tool- making industry that served it on the Birmingham Coventry axis. However, a visit to the city shows wealth and vigour, a clear indicator of the value of a mixed economy and a resilient community – and the replacement of older industries by newer ones. As other authors have observed[7] when writing on the West Midlands, there is a 'tradition of continuous and subtle adjustment' that maintains the city's position in the national and international scene.

ENDNOTES
REFERENCES
INDEX

ENDNOTES

CHAPTER 1

1 Berg 1985.
2 Roe 1916 and 1937, Buckingham 1941. Buckingham's 1941 book is the second edition, the first being published in 1920. Buckingham also wrote an unpublished thesis on the Springfield gauges, personal communication from Peter Smithurst.
3 Howard 1978.
4 Martin Pegler 1998.
5 Blackmore (1961 and 1991, the inspiration for this current work), Roads (1964) and Bailey (1986) on British military firearms. Wilkinson (1841), Greener (1841), Stelle and Harrison (1883), Baker (1928) and Howe (1934/37/41) on the technology of gun making – Stelle Harrison is a manual for traditional craft gun makers, Baker and Howe take a more engineering approach and are aimed at both the professional and the amateur – and Myatt (1979) and Akehurst (1985) on the historical background of guns and gun making especially during the nineteenth and twentieth century. The excellent Goodman, Chairman of BSA (1866), Aitken (1876), Roe (1937), Wise – a key source – (1952), Dunham (undated, usually stated as 1955), Smith – again a key source – (1964), Harris (1949), Allen (1966), Bailey and Nie (1978), Walter (1984), Boothroyd (1986), Tate (1997) and Greener (2000) on the Birmingham gun trade. Deyrup (1948), Gordon (1988 and 1989) and Smithurst (1996, 1998 and 2002) on Armory Practice, Moore (1970) on precision measurement technology and on the key contributions of Rosenberg (1969 and 1976) relating production technology to the growth and contraction of the Birmingham trade. The manufacturing technology focus of the current work is complementary to others (Williams 2000, 2001a, and 2001b, and Williams and Johnson 2001) by the author and colleagues on the technology of gun making.
6 Redford 1960.
7 Crompton 1991.
8 Unfortunately for the wealth of the Black Country, the alternative technology of steel making created in the nineteenth century destroyed the iron-manufacturing industry over 130 years ago (Gale 79). The last blast furnace in the Black Country was blown out in 1977, leaving the Black Country without an operating furnace for the first time in 416 years (Gale 1979). The growth of steel making, however, had the consequence of growing a vigorous steel-based industry in Birmingham.

9 Rolt 1965.
10 Both quotes are from Geier 1949, Geier was one of those who grew Cincinnati Milling Machines and had a deep understanding of the issues!
11 Jackson and de Beer 1973.
12 Roe 1916.
13 Quoted in Hoskins 1951.
14 Gale 1952, 1979.
15 A description of the evolution of the related Birmingham toy trade can be found in Berg 1985. Toys are small metal products usually made in large quantities such as buckles or what we would now call inexpensive or costume jewellery.
16 In the North West of England, the progress of the Industrial Revolution is usually described by the progressive growth of the factory system in the cotton industry and the corresponding evolution of cotton processing machines, machine makers and manufacturing techniques (see for example Musson and Robinson 1960 and Berg 1985).
17 Tholfsen 1952, Behagg 1990.
18 Heaton 1966.
19 Fay 1962.
20 Wood 1976.
21 See also Briggs 1990.
22 Derry and Williams 1960.

CHAPTER 2
1 See Smith 1967 and Tate 1997.
2 Harris 1949 and Williams 2012.
3 Bailey and Nie 1978.
4 See for example Hopkins 1998.
5 Bailey and Nie 1978.
6 Blackmore 1991, Hopkins 1998 indicates the 1760s. Key early suppliers were Cole, Jordan and Farmer
7 Quoted in Blackmore 1961.
8 This 'jigger' was sold in the auction by Christies of the collection of W. Keith Neal on 18 July 2002. The lot description also refers to the only other recorded EIC jigger in D.F. Harding, Small Arms of the East India Co., 1600–1856, Vol.1 pp.207-8.
9 Goodman describes the activities of the Birmingham Tower in 1866:
'When in full operation, a staff of between sixty or seventy men are there engaged in viewing the arms manufactured in the town. The several parts of a gun are first examined in detail, and accurately gauged; they are then returned to the gun maker who proceeds to set them up. At every stage of setting up, the guns are taken for examination to Bagot Street. At each view the examiner strikes his mark on the part examined, and a gun when completed bears twenty two such marks. Each viewers mark has its distinctive number, so that he can be held responsible for any defect which may afterwards be discovered in the subsequent stages of manufacture, or in service.'
10 Or 1798 depending on the source.
11 Walter 1984 or in 1894, again depending on the source.
12 Greener 2000.
13 Blackmore 1991.
14 The Parker Hale 1939 catalogue notes: 'These are useful to assist inexperienced salesmen to ascertain the bore size of shot guns from 4-gauge to 100-gauge'. The gauge of a gun is determined by the size of lead ball required to fit the bore, the gauge is the number of balls to the pound weight of lead. A 12 bore can therefore take a ball of which there are twelve to the pound.
15 See for example Boothroyd 1978 and Newland 2000.
16 Newlands 2000.
17 The lock making business deteriorated at the end of the nineteenth century with the decline in demand for sidelock guns, following the invention of the Anson and Deeley action. The Brazier business was taken over in the 1920s by Edwin Chilton, Chilton's being another well established firm of Wolverhampton lock makers. The Chilton gun lock works went in to liquidation in 1978 (Boothroyd 1978). The only remaining lock maker to the trade in Wolverhampton is the company of York and Wallin Ltd of Bilbrook.
18 Bailey and Nie 1978.
19 Pam 1998.
20 Henderson 1968.
21 Bailey and Nie 1978 identify a Robert P. Little here in 1816–1817.
22 Brumfeld 1985 and 1993.
23 Hopkins 1998.
24 Landes 1969.
25 Henderson 1968.
26 Blackmore 1991.
27 'The products obtained by these dies need only a very small effort to be converted to perfectly finished interchangeable parts.' Gamel 1826. See also Bradley 1990 and Smithurst 1996.
28 Prosser 1881.
29 Smiles 1883.
30 Fitch 1880. Brumfeld 1993 shows some excellent photographs of original flintlock cock die forgings from both the UK and the USA.

31 Woodworth 1916 notes that in the USA, forging was only used after hand working to size for interchangeability. This is what we now call the coining process and is not reflected by other material in this discussion.
32 The top row shows three flintlock forgings and two hand forgings for percussion; the lower row shows five centre fire hammers and two percussion hammers.
33 Akehurst 1985.
34 Stelle and Harrison 1883 and Brumfeld 1985.
35 Encylopedie, ou Dictionarre Raisonne des Sciences, des Arts et des Metier, 1751–65, Denis Diderot, Paris, see Brumfeld 1985.
36 Howe 1941 gives detailed instructions on how to make a cherry.
37 There is some evidence remaining of one of these water powered sites, 'The Birmingham Gun Barrel Works' on the Stour at Hayseech, Old Hill, SO 959850, where some of the dam is visible (Association for Industrial Archaeology 1991).
38 Tate 1997.
39 Tate 1997.
40 Behagg 1998.
41 Translated in full in Henderson 1968.
42 Akehurst 1985.
43 London making of twist barrels had stopped in 1833 with the death of W. Fullerd (Akehurst 1985).
44 Anon 1881.
45 Blackmore 1991.
46 Court 1929.
47 Habakkuk 1962.
48 Prosser 1881 and Bailey and Nie 1978.
49 This John Jones is usually considered to be the John Jones who went to Tula in Russia with his son, also John Jones.
50 Behagg 1998.
51 Goodman also tells us that in the American Civil War there was such a large demand for military barrels that twist barrels were supplied. The twisted barrels withstood proof somewhat better than the plain iron ones 'twisted 1.20 as compared with plain iron 1.35 in the hundred; yet the difficulty of obtaining them as perfectly free from flaws and specks as is required for a rifle barrel led to their rejection as soon as rolled barrels could be procured in sufficient quantities'.
52 Harris 1949.
53 George 1962.
54 Smith 1967 and Williams 2010.
55 Goodman 1866.
56 Data collated by Fries 1975 from the public record office on British firearms exports for 1844 shows that 164,251 muskets of a total of 337,567 were exported to Africa or West Africa and had an average value of £0.47. This contrasts to the 41,466 and 44,888 muskets exported to India and Asia with a value of £2.2 and £2.10 respectively.
57 George 1947.
58 Rolt 1965.
59 Also famous for his Calculating Engine.
60 Behagg 1998.
61 Behagg 1998.
62 Berg 1985.
63 Henderson 1968 and Berg 1985.
64 Davis 1973 and Boaz 1996.
65 Goodman 1866.
66 Winant 1961.
67 Swenson 1971.
68 Winant 1961.
69 Harris 1949.
70 Harris 1949.
71 Bailey and Nie 1978.
72 Bailey and Nie 1978.
73 Greener 2000.
74 Goodman 1866.
75 Bailey and Nie 1978.
76 From de Witt Bailey 1997.
77 Moore 1970.
78 Wise 1952 and Smith 1964.
79 The Smith map shows the halving of the area occupied by the trade between 1948 and 1964, the construction of Queensway, the Birmingham Ring Road halved the size of the trade again in the 1970s. The trade is now largely confined to Price Street.
80 This pattern was further emphasised as the gun trade subsequently contracted, small businesses and individual craftsmen moved into the Victorian factories built in the expansion of the trade during the American Civil War for example.
81 An outworker is someone who carries out a manufacturing task for an entrepreneur, usually from their own home or sometimes workshop.
82 Rosenberg 1969.
83 Rosenberg 1969.
84 Identified in Hopkins 1998.

CHAPTER 3
1 Kirkland 1994 and others.
2 Blackmore 1991.
3 Kirkland 1990.
4 Roe 1916 and Woodbury 1977 who notes that Le Blanc's techniques are described in full in his Memoire of 1790. See Alder 1997 for a full account.
5 George 1947.

6. Blackmore 1961. Williams 2016. Key early suppliers were Cole, Jordan and Farmer.
7. Battison 1988 introducing Gamel 1826.
8. Roe 1916.
9. Swenson 1971.
10. Anon 1868.
11. The anonymous writer of this notes that Bodmer was also driven by the needs of the Napoleonic Wars for firearms but that 'As the percussion lock was not known, Mr Bodmer had, of course, to contend with great difficulties; and it is scarcely necessary to observe that the principle was never adopted, the time for a change of such magnitude having evidently not arrived'.
12. Roe 1916.
13. From Anderson 1858. Anderson understood the principle clearly; the paper from which the figure is taken describes how he has applied the approach at the Woolwich Arsenal.
14. Blackmore 1991.
15. Ames and Rosenberg 1970.
16. Fries 1975.
17. Fries 1975.
18. A similar machine is on display in the Science Museum in London, as is an Ames inletting machine.
19. Cooper 1991 and 1988.
20. Deyrup 1948.
21. Rosenberg 1976.
22. Battison 1966 shows that the Whitney locks of his first batch of muskets are marked with a number and are therefore not interchangeable.
23. Habbakuk 1962, Uselding 1970 and Gordon 1983, for example.
24. Woodbury 1960.
25. Gordon 1997.
26. Roe 1916.
27. Swenson 1971.
28. Schmidt 1996.
29. Roe 1916.
30. Rolt 1965.
31. Gordon 1997.
32. Smithurst 2002.
33. It is important to recognise (Roe 1916) that it was expected that when supplying weapons for government contracts 38 per cent of them would be rejected – but that these second rate rifles could be sold off. This highlights the problems associated with achieving the interchangeability ambitions of the military.
34. Rosenberg 1969.
35. Roe 1916.
36. Rosenberg 1976.
37. Callagahn 1972.
38. Howard 1978.
39. Woodbury 1959.
40. Initiated by Battison 1966.
41. Merrit Roe Smith (1973 and 1977) and Rosenberg (1969). The excellence of Rosenberg's contribution to the description of interchangeable manufacturing and technology transfer from the USA to UK and vice versa is identified by Hounsell 1984 in his wider study of the achievement of interchangeablity in industries outside gun making.
42. Deyrup 1948.
43. Hounsell 1984.
44. Howard 1980.
45. Howard 1978.
46. Smith 1977.
47. Deyrup 1948.
48. Schmidt 1996 gives an excellent description of Hall's rifles.
49. Battison 1988, confirming the importance of Hall.
50. Further descriptions of this and the later transfer of USA technology via the UK and the machine maker Thomas Greenwood to Tula are made in Joseph Bradley, *Guns for the Tsar, American Technology and the Small Arms Industry in Nineteenth Century Russia*, Northern Illinois Press, 1990.
51. See the description of craft flintlock making (Brumfield 1985).
52. Merritt Roe Smith 1973 and 1977.
53. Henry Maudslay also used metal in many of his machine tools, for example all but two of the machines in the Portsmouth block-making machinery completed in 1809 were constructed in metal (Gilbert 1971).
54. Fitch 1880.
55. Merritt Roe Smith 1977.
56. Burton was responsible for transferring many of the USA techniques to Europe and Russia.
57. It is interesting to compare this with techniques being applied in the UK in other gunmaking centres. Birmingham was also the centre of the steam engine and toy trades and many manual finishing processes were used. Machine tool making was at other UK centres, but as is tradi-tional, made little pretence of being high volume or interchangeable. It is clear that Hall was one of the first to take an engineering approach to the whole problem. This contrasts, for example, with the early days of the steam engine; Watt was essentially an engineering scientist who schemed his engines as drawings, Boulton was an entrepreneur – the design of the manufacturing process was essentially left to the pattern maker

– and artisan – who detailed the design including with, for example, draft angles and radii to allow manufacturing as he constructed the patterns.
58 Gordon 1989.
59 See **2.1** auctioned at Christies, London in July 2002, now in the Royal Armouries, Leeds.
60 Gauge supplied to the East India Co. by Ezekial Baker, 1818, illustrated in Smallarms of the East India Co. 1600-1856 Vol.1, Procurement and Design, D.F. Harding, 1997.
61 Moore 1970.
62 Roe 1916.
63 Gordon 1988b.
64 Deyrup 1948.
65 Gordon 1997.
66 Colvin 1913.
67 Atkinson 1996 and Williams 2007.
68 Roe 1916.
69 The limit gauge or equivalent dimensional measurement identifies that the component is within a specified tolerance range usually denoted as plus or minus, +/-, a nominal dimension. Fit and function is determined by the relation of the mean dimension and tolerance ranges of the components that interact and is determined by the choice of dimension, tolerance range 'limits' and the characteristics and function of the 'fit' required, for example interference, sliding or clearance. The engineering practice of limits and fits is much later than the discussion here. J.W. Newall published the first standard tables of limits and fits in 1902 with the first British Standard, No.27, being published for these in 1907 (Koenigsberger 1978). By the adoption of this and the later standard of 1924: 'vagueness is abolished and accurate interchangeability is made possible' (HMSO 1940). German and American Standards followed in 1921 and 1924 respectively (Koenigsberger 1978). Modern practice also considers the statistical likelihood of the chosen manufacturing process being able to achieve a dimensional tolerance – process capability. Natural tolerances were in post-war practice chosen to be equal to plus or minus three standard deviations of the population of dimensions achieved by the process (Shigley 1972). This meant that 99.73 per cent of production was within the tolerance limits and there-fore required inspection or other steps to filter the 2,700 parts in a million that would be outside the acceptable tolerance range. Industry also practices 'selective assembly' in machine and machine element building where components are made within a tolerance range and then measured to determine their position within the tolerance so that they can be matched to mating components selected to give appropriate fits. Recent 'six sigma' approaches to quality (Breyfogle 1999) such as those pioneered by Motorola, require capable manufacturing processes to be chosen that match the requirement such that plus or minus six standard deviations of the output of the process are within the tolerance limits for critical dimensions or characteristics even with process drift. This gives 3.4 parts per million failure rate for each process step and can eliminate the need for inspection, thereby improving the overall economics of the manufacturing system.
70 Parsons 1947.
71 Habakkuk 1962.
72 Landes 1969, Saul 1967.
73 Williamson 1972.
74 Cooper 1991.
75 Deyrup 1948.
76 F.W. Taylor was one of the pioneers of production engineering and the inventor of high speed steel. He showed that the life of a cutting tool could be described by a mathematical model: Taylor's tool life equation. He captured and disseminated the principles of Scientific Management, a scientific approach to the management of factories that included the technique of work measurement – 'time and motion'. This led to many tensions between capital and labour.
77 Woodbury 1960.
78 See for example Battison 1973.
79 If this machine was from Harpers Ferry, examination of the Hall rifle suggests that the slide could have been required to allow 8 to 10in of movement in order to make some of the large prismatic features.
80 For example Gordon1988 a and b and 1989.
81 Malone 1988.
82 Gordon 1988a.
83 Gordon 1988b.
84 Malone 1988.
85 See for example Raber 1988.
86 Fries 1975.

CHAPTER 4

1 At this time the Institution of Civil Engineers was the senior UK civilian (opposed to military) engineering institution, being founded in 1818 'for the general advancement of mechanical science'. Over the years it began to focus more and more on what is now understood as civil

engineering, for example the construction of road systems, bridges and other static works. The 'Mechanicals' was founded in Birmingham in 1847 to provide a forum for those primarily interested in the manufacture and operation of moving machinery. There was also some ongoing professional rivalry between individuals. George Stevenson was unwilling, given his professional standing, to write the probationary essay deemed a necessary condition 'as proof of his capacity as an engineer' of his membership of the 'Civils' – and started with others the rival institution, becoming its first president! (Parsons 1947).
2. Blackmore 1970.
3. Fries 1975, who suggests this is a fourfold saving over hand manufacture.
4. Cogan 1999 when writing on materials cautions us on interpreting the language used in early publications with the technical language we use today – the use of the phrases 'forging and swage' echoes that we have a similar problem when trying to establish the detail of the metal-forming processes used by our earlier colleagues. W.W Greener (1910) describes a closed die forging as 'a stamping' and a hammer forging as 'a forging'.
5. Conducted by Taylerson et al. (1966) and others particularly Sellars and Seymour see Howard 1978.
6. Grant 1995.
7. Wilson also notes in 1978: 'There are machines still in use at Colt's that date back to the time of E.K. Root and Samuel Colt'.
8. The hand is the finger that indexes the revolver cylinder by pushing on a ratchet machined on the cylinder.
9. Howard 1978.
10. Two are shown in Rosa 1988.
11. Rosa 1988.
12. Rosenberg 1969.
13. Form cutters.
14. J.N. George (see the 1962 edition of his book), the collector historian describes 'cheap, nameless, Birmingham-made revolvers of the middle fifties' as follows: 'Their quality was in general of the very poorest... on the rough and ready principle that if the finished revolver would fire its five proof rounds, without bursting or breaking down, it would do well enough for the type of customer who simply demanded "a revolver" for 10s or 15s, without specifying that it should shoot straight'.
15. Taylerson et al. 1966.
16. See also Rosa 1988.
17. Jackson et al. 1989.
18. Taylerson 1968.
19. Blackmore 1970.
20. The Colt Armory at Hartford was completed in 1855. Root took over presidency of the business following Colt's death in 1862. Roe 1916 shows illustrations of Root's splining machine and multi head chucking lathe of about 1855. Root himself died in 1865. Much of the original machinery in the factory was destroyed by a disastrous fire in 1864. Colt's widow funded rebuilding and re-equipping of the factory. Colt employed Alexander Thuer, an English gun maker as a contractor to lead his hammer and hammerless shotgun making. Colt used the system of internal contractors as discussed earlier in the book. In the 1940s it was noted that the machinery in the factory was little changed after a 100 years – Colt had very significant financial problems during the Second World War, in spite of the fact that conflict is usually advantageous to the arms trade. Colt finally moved out of the Armory in 1994. Spencer, Billings, Pratt & Whitney all spent time at Colt.
21. Gilbert 1960.
22. Dunham 1955.
23. Atkinson 1996.
24. According to Atkinson 1996.
25. Goodman 1866 describes this well: 'The process of "setting" a barrel, that is straightening it, is another branch calling for the very highest skill. It must be understood that the degree of straightness required in a gun barrel is nothing short of absolute perfection. The practised eye of a barrel setter can detect a deviation from the straight line which no mechanical contrivance can discover. He accomplishes his object by looking through the barrel, while standing in front of a window, and causing the shade of the upper edge of the window to traverse up and down the tube. The irregularities in the outline of the shade show him where the inaccuracies exist. These he removes by well directed blows of a hammer, the perfecting blows being given by a light wooden mallet.' This practise is still used today, the mallet being replaced by a screw jack. It relies on the specular finish of the barrel interior for success. Goodman adds the following footnote: 'We learn from a MSS account of the gun manufacture, written by Mr Hawkes Smith some thirty years ago, which has been kindly placed in our hands by Mr Toulmin Smith, that this mode of testing

the accuracy of the inside of a barrel was the discovery of a Birmingham workman (a fine borer) forty or fifty years before. The discovery, which Mr Smith justly estimates as worth many thousand pounds, was sold by the shortsighted discoverer for five guineas and a pot of ale.'
26 Whitworth had introduced first angle drawings into his factories by then – this technology clearly had not reached the gun trade.
27 The original Victorian photographs of the Ames and Robbins & Lawrence Machinery show in this and the previous chapter are thought by Roads 1965 to be of the machines installed by the London Armoury Co.
28 Gilbert 1963.
29 Gilbert 1963.
30 Swenson 1971.
31 Gilbert 1960.
32 Blackmore 1961.
33 Roads 1961.
34 Paragraph 116 of the Select Committee Report.
35 Pam 1998. Blackmore gives an excellent account of the growth of mechanisation at Enfield in his 1991 paper.
36 Blackmore 1991.
37 Enfield closed in 1988 and the Pattern Room relocated to the Royal Ordnance Factory Nottingham. This has now closed and the Pattern Room has been further relocated to the Royal Armouries Leeds.
38 See Prosser 1881.
39 Bailey, de Witt 1986.
40 Blackmore 1991.
41 Moore 1970.
42 Rosenberg 69.
43 Jacques 1993.
44 Bailey, de Witt 1986.
45 See Taylerson 1966.
46 A second London Armoury Co. was founded in 1894 (Taylerson 1966).
47 Smith 1977.
48 Smith 1974.
49 Gilbert 1963.
50 Mordecai 1860.
51 Goodman 1866.
52 Derry and Williams 1960.
53 Anon 1890.
54 Donaldson 1909 quoted in Landes 1969.
55 Roads 1964.
56 Smiles 1883 and Moore 1970.
57 Browne & Sharpe also introduced the French invented hand-held micrometer at this time.

It was subsequently introduced into the UK by Charles Churchill in around 1900. Other significant steps in measurement techniques were the introduction of the dial gauge in about 1920, the mechanical comparator in 1930, the electronic comparator in 1948 and the Talysurf to measure surface finish in 1940 (Koenigsberger 1978).
58 Moore 1970.
59 Cantrell and Cookson 2002 give an excellent up to date account of early UK machine tool technology.
60 Gilbert 1960.
61 Rosenberg 1969.
62 Habakkuk 1962.
63 Rosenberg 1969.
64 Cogan 1999.
65 Moore 1970.
66 Whitworth 1873.
67 Heaton 1966.
68 Rosenberg 1969.
69 Fries 1975.
70 See for example Ames and Rosenberg 1970.
71 Allen 1929 and Behagg 1998.
72 Goodman 1866.
73 Richard B. Prosser was the son of Richard Prosser who gave evidence to the committee of 1854.

CHAPTER 5
1 Walker 1984. See Williams 2013 on Greener.
2 Blackmore 1961, Roads 1964.
3 Roads 1964.
4 Current rifle barrels were of course 39in long!
5 Letter to *The Times*, April 29 1857.
6 This and what follows is largely drawn from Walter 1984.
7 Some say fourteen.
8 Note that neither Greener or Westley Richards figure in these names.
9 See Tate 1997. Bentley & Playfair's premises were one of those occupied by small businesses when the company failed.
10 Bailey and Nie 1978. See Williams 2013 on Goodman.
11 He talked of the making of 'good sound pistols, because every part is made by machinery; they are all sound; there is no high finish about them'. Unfortunately, later he shows that he does not understand the detail of the factors that affect the quality of work saying: 'I possess a mercantile knowledge of these matters, but beyond that I do not profess to know'. He also talks of the difficulties of obtaining machine-made articles:

'there are some machine-made articles that we have greater difficulty in procuring than any others; one that is made wholly by machinery we have the greatest possible difficulty in, that is the article of wood screws in which every process is conducted by machinery'. As an apprentice in GKN in Smethwick on the Birmingham Black Country boundary in the early 1970s, the largest maker of fasteners and wood screws in the world at that time, I remember seeing wood screw machines from the 1870s that still ran occasionally. Echoing the experience in gun making, the techniques used in these machines were acquired when Nettlefold (the N in GKN) and Chamberlain, subsequently Mayor of Birmingham, an MP and Cabinet Minister, bought patent rights to American machines in 1854 (Saul 1967).

12 Greenwood also supplied many machines to Enfield.
13 Goodman 1866.
14 This machine is clearly to the drawing shown in **5.14**. The level of finish of the Greenwood machine is not high when compared with the Ames machine shown in **4.1** and the Ames machine held in the Science Museum illustrated in Gilbert 1966. While the Greenwood machine has been working for sixty years and the Ames machines are new and restored respectively, there are differences in finish. The superior finish of the Ames machine challenges the view of American machine makers taken by the British at Colts' lecture to the Institution of Civil Engineers. We should also not be surprised that the machine was being used to make the SMLE in 1917; many of the original Ames machines in Springfield were also being used in to manufacture the Model 1903 Springfield in the Great War (Raber 1988). The major technical steps in the early years of the twentieth century were 'little kinks and devices' – the use of special purpose jigs, fixtures and machine components to turn adapt older more general purpose machines into an effective production system (Malone 1988).
15 These gauges include some for motorcycle frame components.
16 Goodman 1866.
17 Fries 1975.
18 Goodman 1866.
19 Anon 1979.
20 Greener 1871.
21 Fries 1975.
22 Belton 1984 or in 1887 (Pam 1997).
23 HMSO 1887.
24 Anon 1979.
25 Taylerson 1966.
26 Proof House, undated. Accles & Pollock was subsequently part of the TI Group.
27 Greener 2000.
28 Lewis 1996.
29 Lewis 1996, Saul 1967.
30 Atkinson 1996
31 Floud 1976, Smithurst 1998
32 Bedford 1984.
33 Swenson 1971, Bradley 1990.
34 In Bradley 1990, we find the following quote from the Russian Inspector of Small Arms, Captain Bil'derling: 'The machine production I saw in America is a long way from acceptance here in England where hand production reigns supreme. The factory here is just as unfamiliar with the small calibre rifle as we are. They admit that they have never made small-calibre arms and consequently had to learn on our order. Having learned at our expense, at the end of the order they will cease being accommodating and will require us to pay a huge amount for the patterns, tools and machines, which in fact we specified. As for the quality of English workmanship it is much lower than American. The shocking percentage of defects in the parts has barely decreased: 50 per cent of the barrels and stocks, 30 per cent of the frames, 25 per cent of the mechanisms, and 15 per cent of the ramrods are defective.'
35 Greenwood 1862.
36 The distinguished Scottish engineer William Fairbairn notes a connection with the manufacture of the copying machinery for wood processing at Woolwich in the discussion following the delivery of Anderson's 1858 paper.
37 Greenwood 1868.
38 Battison in his introduction to Gamel 1826.
39 Greenwood 1868.
40 Blackmore 1965.
41 Today we have other mechanised solutions to remove such repetitive tasks, for example the bowl feeder.
42 Moore et al. 1921.
43 Grimston 1928.
44 Labbett 1984.
45 Quoted in Tate 1997.
46 Again quoted in Tate 1997.
47 Anon 1881.
48 Horn shavings and burnt leather from old shoes were also used. Akehurst (1985) says: 'The iron

at red heat absorbed the carbon and on being cooled the surface became very hard and at the same time took on the mottled colouring that was so desirable. Iron in this state could still be bent or set as the hardening was on the surface only'.
49 Dowell 1962.
50 Saul 1967.
51 These photographs are taken from the folio Archdale 1890 held by the Birmingham Proof House.
52 Floud 1976.
53 Nie and Bailey 1981.
54 Taylerson 1966.
55 Anon 1962.
56 Walker 1984.
57 Allen (1966) writing in 1929 highlights that: 'In spite of the prosperity of the military section, however, it seems that even in 1860 the sporting gun section branch was larger'.
58 Judd 1993.
59 Habakkuk 1962.
60 Shown in Bailey and Nie 1978.
61 Taylerson 1966.
62 Howe 1941.
63 Howard 1978 and 1980.
64 Goodman 1866.
65 That by Howe 1941 is however excellent.
66 Roberts 1952.
67 Blackmore 1961.
68 Rosenburg 1969.
69 Walsh 1884.
70 Pam 1997.
71 Kolbe 1991 and Heard 1997.
72 ASM 1969.
73 Rosenberg 1976.
74 Rosenberg 1976.
75 Roe 1916 and Woodworth 1916.
76 Schelsinger 1941.
77 Rolt 1965.
78 Anon 1910.
79 Woodworth 1916.
80 Subsequently the motorcycle and the motor car. While gun making introduced precision machining, gauging and some fitting, the technique of selective assembly came from watch making.
81 See Moore 1970.
82 The 'Swedish Mauser' is considered by target shooters to be one of the most accurate military rifles, Kehaya 1999, this may be because of the emphasis on precision from Johansson.
83 Rosenberg 1976.
84 Goodrich and Stanley 1912.
85 Moore 1970.
86 Saul 1967 sees Remington – cycles, typewriters etc. – as the benchmark for this.
87 Saul 1967.

CHAPTER 6

1 Akehurst 1985.
2 1910 edition.
3 Fries 1975.
4 Taylor 1913.
5 The choke is the constriction machined into a shotgun barrel to control the shot pattern.
6 Quoted in Akehurst 1985.
7 Now the site of the Merry Hill shopping centre.
8 The chopper lump integrates into a single component the barrel tube and the lump that locks the barrel into the rest of the action.
9 There is an excellent description of the need for fitting in *The Art of Shooting* by Charles Lancaster, originally published in 1898 (see Lancaster 1954 for example). Lancaster also, interestingly for the context of this book, comments on the definition of bore as follows: 'The term "bore" used to indicate calibre, is a survival from times when facilities for fine measurements were not generally available, and in the making of firearms, were not greatly needed'.
10 Fries 1975.
11 Taylerson 1970.
12 Dunham 1955.
13 Allen 1966.
14 Harris 1949.
15 Wise 1948.
16 Harrison 1978.
17 Both Tate 1997 and Boothroyd 1986 include material on the Carrs, and fortuitously Smith 1964 gives a full description of their workshops. Further, two of their day books are in the author's possession.
18 Harris 1949.
19 Described by Smith 1964.
20 Guns were sometimes test fired by pushing them through the window, the sound of pellets falling on the roof being an everyday experience!
21 In full in Williams 2002 – Bill Roper wrote to the author after the publication of this piece.
22 Recorded by Wise in 1948, Smith 1964.
23 Tate 1997 and Boothroyd 1986.
24 *The Ideal Gunworks* of Thomas Chambers, see **2.34**, operated from 15 St Mary's Row during this period and became part of William Ford for a period.
25 Smith 1964.

26 These now come from Belgium.
27 Boothroyd 1970.
28 The original photograph is reproduced in Greener 2000. Graham Greener's book also shows Charlie Kyte at work in Greener's in the 1950s.
29 These workshops are some of the most photographed in Birmingham. As the reader will notice, they appear a number of times in this book and in others including the definitive history of Victorian Birmingham, Briggs 1952. They also appear a number of times in the series of articles by Boothroyd 1969.
30 Boothroyd 1969.
31 Greener 2000.
32 Westley Richards factory moved to Bournbrook in 1898. From 1812 to 1898 they occupied 82 High Street Birmingham.

CONCLUSIONS

1 Roe 1916.
2 see for example Berg 1985.
3 Bailey and Nie 1978.
4 See the discussion in Fries 1975.
5 Howe 1941 etc.
6 Westley Richards 2000.
7 Wood 1976.

REFERENCES

Aitken W.C., *The Birmingham Trades: Guns, in British Manufacturing Industries*, Edward Stanford, 1876.
Akehurst R., *Game Guns and Rifles: Percussion to Hammerless Ejector in Britain, Arms and Armour Press*, 1985.
Alder K., *Engineering the Revolution, Arms and The Enlightenment in France, 1763–1815*, Princeton University Press, 1997.
Allen G.C., *The industrial development of Birmingham and the Black Country 1860–1927*, Frank Cass, 1966 (Reprint).
American Society for Metals, Metals Handbook, Vol.4, Forming, 1969.
Ames E. and Rosenberg N., 'The Enfield Arsenal in Theory and History', in Ed Saul S.B., *Technological Change: The United States and Britain in the Nineteenth Century*, 1970.
Anderson J., 'On some applications of the copying or transfer principle in the production on [sic)]wooden articles', *Proceedings of the Institution of Mechanical Engineers*, 1858, Vol.9, 237-248.
Anon, 'Birmingham Small Arms Co., Small Heath Birmingham, and Messrs Webley and Scott, Gun Manufacturers Birmingham', *Proceedings of the Institution of Mechanical Engineers*, 1910, Vol.58, Part 3–4, 1324-1327 and 1356-1357.
Anon, 'British Industries, No CXIV, Messrs P. Webley and Son, Gun Rifle and Revolver Manufacturers, Birmingham', *British Trade Journal*, 1 October 1881.
Anon, *Memoirs, Proceedings of the Institution of Civil Engineers*, Vol.28, 1868, p.573.
Anon, 'The March of the Piled Arms', un-published manuscript, 'The Birmingham Gun Barrel Proof House', PRO/70/58 and May 1979.
Anon, *Under Five Flags, the Story of the Kynoch Works*, Witton, Birmingham 1862–1962, 1962.
Archdale, J., Folio of Photographs in the Birmingham Gun Barrel Proof House Collection, 1890.
Association for Industrial Archaeology, *Industrial Archaeology of the West Midland Iron District*, 1991.
Atkinson N., Sir Joseph Whitworth *The World's Best Mechanician*, Sutton Publishing, 1996.
Babbage C., *Economy of Manufactures*, Second Edition, 1832?
Baker C., *Modern Gunsmithing*, Small Arms Technical Publishing Co., 1928, Facsimilie Border Press, 1995.
Baker E., *Remarks on Rifled Guns*, 1832 facsimilie by Standard Publications Inc, Huntington, West Virginia.
Bailey De Witt, *British Military Longarms 1715–1865*. London: Arms and Armour Press, 1986.
Bailey De Witt, *Pattern Dates for British Ordinance Small Arms 1718-1783*, Thomas Publications, 1997.

Bailey D.W. and Nie D.A., *English Gunmakers, The Birmingham and Provincial Gun Trade in the eighteenth and nineteenth century*, Arms and Armour Press, 1978.

Battison E.A., 'Eli Whitney and the Milling Machine', *The Smithsonian Journal of History*, Vol.1, No.2, Summer 1966, pp 9–34.

Battison E.A., 'A new look at the 'Whitney' Milling Machine', *Technology and Culture*, 14, 592–598, 1973.

Beaufroy H., Scloppetaria: *Or considerations on the nature and use of rifled barrel guns*, 1808.

Bedford D., 'Restoration of a rifling machine', *Black Powder*, Vol.31, 1984, pp.32–33.

Bedford D., 'Restoration of a rifling machine – Part 2', *Black Powder* Vol.32, 1985 pp.12–13.

Behagg C., *Politics and Production in the Early Nineteenth Century*, Routledge, 1990.

Behagg C., 'Mass production without the factory: Craft producers, Guns and Small Firm Innovation, 1790-1815', *Business History*, 40, No.3, pp.1–15, 1998.

Belton J.A., 'Edward Jones' Patent, and the Westley Richards Co., Ltd', *Arms and Arms Collecting*, 22, No.4, 128–129,1984.

Blackmore H.L., *British Military Firearms 1650–1850*, Herbert Jenkins, 1961.

Blackmore, H.L., *Guns and Rifles of the World*, Chancellor Press, 1965.

Blackmore H.L., 'Colt's London Armoury', in Ed Saul S.B., *Technological Change: The United States and Britain in the Nineteenth Century*, 1970.

Blackmore H.L., 'Military Gun Manufacture in London and the Adoption of Interchangeability', *Arms Collecting*, Vol.29, No.4, Nov 1991, pp.111–122.

Berg M., *The Age of Manufactures*, 1700-1820, Fontana, 1985.

Boaz T., *Guns for Cotton, England arms the Confederacy*, Burd Street Press, 1996.

Boothroyd G., *The Handgun*, Cassell, 1970.

Boothroyd G., 'The Bimingham Gun Trade, A series of six articles', *Shooting Times and Country Magazine*, October–November 1969.

Boothroyd G., 'Wolverhampton Gun Locks', *Shooting Times and Country Magazine*, March 2–8, 1978, pp.30–31.

Boothroyd G., *Shotguns and Gunsmiths, the Vintage Years*, A.C. Black, 1986.

Bradley J., *Guns for the Tsar, American Technology and the Small Arms Industry in Nineteenth Century Russia*, Northern Illinois University Press, 1990.

Breyfogle F.W., *Implementing six sigma: Smarter solutions using statistical methods*, Wiley, 1999.

Briggs, A., *History of Birmingham, Vol.II: Borough and City 1865–1938*, Oxford University Press, 1952.

Briggs, A., *Victorian Cities*, Penguin Books, 1990.

Brown & Sharpe, *A Practical Treatise on Milling and Milling Machines*, Providence, RI, 1914.

Brumfield, G., 'The production of flintlocks used in colonial American rifles: Raw Materials, Tools and Technology', *Journal of Historic Arms Making Technology*, Vol.1, pp.1–81, 1985.

Brumfield, G., 'Cock Forging, A Study in Technology', *Journal of Historic Arms Making Technology*, Vol.5, pp.65–87, 1993.

Buckingham, E., *Interchangeable Manufacturing*, Second Edition, The Industrial Press, 1941.

Chambers J.D., *The Workshop of the World, British Economic History 1820–1880*. Oxford: Oxford University Press, 1974.

Cogan A., *Fighting Iron: A metals handbook for arms collectors*, Andrew Mowbrey, 1999.

Callaghan B.A., 'The Hall Carbine Scandal', *Guns Review*, 12, No.8, 300–302, 1972.

Cantrell J. and Cookson G., *Henry Maudsley and the Pioneers of the Machine Age*, Tempus, 2002.

Colt S., 'On the application of Machinery to the manufacture of Rotating Chambered-Breech Fire Arms, and the peculiarities of those arms'. *Minutes of the Proceedings of the Institution of Civil Engineers*, Vol.9, 30-68.

Colvin F.H. and Hass L.I., *Jigs and Fixtures*, McGraw-Hill, 1913.

Cooper C.C., 'A Whole Batallion of Stockers: Thomas Blanchard's Production Line and Hand Labour at Springfield Armory', *Industrial Archeology*, Vol.14, No.1, 1988, 37–58.

Cooper C.C., *Shaping Invention, Thomas Blanchard's Machinery and Patent Management in Nineteenth Century America*, Columbia University Press, 1991.

Court W.H.B., *The rise of the Midland Industries, 1600–1838*, Oxford University Press, 1952.

Crompton J., *A Guide to the Industrial Archaeology of the West Midlands Iron District*, Association for Industrial Archaeology, 1991.

Davis C.L., *Arming the Union, Small Arms in the Civil War*, Kennikat Press, 1973.

REFERENCES

Deane's Manual of the History and Science of Firearms, Longman, Brown, Green, Longman and Roberts, 1858, Facsimile Standard Publications Inc.

Derry T.K and Williams T.I., *A Short History of Technology*, Oxford University Press, 1960.

Deyrup F.J., *Arms Making in the Connecticut Valley*, 1948 (in reprint).

Dowell W.C., *The Webley Story, 1962*, Commonwealth Heritage Foundation, Reprint 1987.

Dunham K., *The Gun Trade of Birmingham*. Birmingham: City of Birmingham Museum and Art Gallery, undated.

Fay C.R., *Great Britain from Adam Smith to the Present Day*, Longmans, 1962.

Fitch C.H., *Report on the Manufactures of interchangeable parts, Tenth Census of the United States*, II, 14, 1880 (quoted in Roe 1916, Rosenberg 1969).

Floud R., *The British Machine Tool Industry, 1850–1914*, Cambridge University Press, 1976.

Fries R.I., 'British Response to the American System: The case of the Small-Arms Industry after 1850', *Technology and Culture*, Vol.16, 1975, pp.377–403.

Gale W.K.V., *Boulton, Watt and the Soho Undertakings*, City of Birmingham Museum and Art Gallery, Department of Science and Industry, 1952.

Gale W.K.V., *The Black Country Iron Industry*, The Metals Society London, 1979.

Gamel I., *Description of the Tula Weapon Factory in Regard to Historical and Technical Aspects, 1826*, translation edited and introduced by Battison, E.A, Amerind Publishing, 1988.

Geier F.V., The *Coming of the Machine Tool Age – the tool builders of Cincinnati, The Newcomen Society of England*, New York Branch, 1949.

George J.N., *English Guns and Rifles*, Small Arms Technical Publishing Co., 1947.

George J.N., *English Pistols and Revolvers*, Second Edition, The Holland Press, 1962.

Gilbert K.R., *Science Museum Machine Tool Collection, Catalogue of Exhibits with Historical Introduction*, HMSO, 1960.

Gilbert K.R., 'The Ames recessing machine: A survivor of the original Enfield Rifle Machinery', *Technology and Culture*, Vol.4, 1964, 207–213.

Gilbert K.R., *Henry Maudsley, Machine builder*, Science Museum booklet, HMSO, 1971.

Gordon R.B., 'Materials for manufacturing: The response of the Connecticut Iron Industry to Technological Change and Limited Resources', *Technology and Culture*, Vol.24, 1983, pp.602–634.

Gordon R.B., 'Who turned the mechanical ideal into mechanical reality', in *Technology and American History*, ed Cutcliffe S.H. and Reynolds T.S., University of Chicago Press, 1997.

Gordon R.B. 1988a, 'Material Evidence of the Manufacturing Methods used in "Armory Practice"' *Journal of the Society for Industrial Archeology, Theme Issue: Springfield Armory*, Vol.14, pp.23–35, 1988.

Gordon R.B. 1988b, 'Gaging, Measurement and the Control of Artificer's Work in Manufacturing, *Polhem*, Vol.6, pp.159–172, 1988.

Gordon R.B., 'Simeon North, John Hall and Mechanised Manufacturing', *Technology and Culture*, Vol.30, pp.179-188, 1989.

Goodman J.D., 'The Birmingham Small Gun Trade' in Samuel Timmins, ed *Birmingham and The Midland Hardware District*, London 1866.

Goodrich C.L. and Stanley F.A., *Accurate Tool Work*, McGraw-Hill, 1912.

Grant E.S., *The Colt Armory, A History of the Colt Manufacturing Company*, Inc., Mowbrey, 1995.

Greener G., *The Greener Story, The History of The Greener Gunmakers and their Guns*, Quiller Press, 2000.

Greener W., *The Science of Gunnery as applied to the use and construction of firearms*, Longman, 1841.

Greener W.W., *The Gun and its Development*. 9th edition 1910. Facsimile New York: Bonanza Books, 1967.

Greener W.W., *Modern Breech loaders*, 1871, Reprint by LUJAC Publishers, Colorado, undated.

Greenwood T., 'On Machinery for the manufacture of gunstocks', *Proceedings of the Institution of Mechanical Engineers*, Vol.13, 1862, pp.328–34.

Greenwood T., .Description of the Machinery for the manufacture of the Boxer cartridges., *Proceedings of the Institution of Mechanical Engineers*, Vol.19, 1868, pp.105–134.

Grimston, F.S., 'Season-cracking of small arms cartridge cases during manufacture', *Journal of the Institute of Metals*, Vol.39, p 255–278, 1928.

Habbakkuk H.J., *American and British Technology in the Nineteenth Century: The search for labour saving inventions*, Cambridge University Press, 1962.
Harris C., *The History of the Birmingham Gun Barrel Proof House*, The Guardians, 1949.
Harrison D., *Salute to Snow Hill*, Barbryn, 1978.
Hawker P., *Instructions to young sportsmen in all that related to guns and shooting*, Longman, 1838, 9th edition.
Heard B.J., *The handbook of Firearms and Ballistics: Examining and Interpreting Forensic Evidence*, Wiley, 1997.
Heaton H., *Economic History of Europe*, Harper International, 1966.
HMSO, Her Majesty's Stationery Office, *Report of the Committee appointed to enquire into the Organisation and Administration of the Manufacturing Departments of the Army*, Eyre and Spottiswoode, 1887.
HMSO, His Majesty's Stationery Office, *Text Book of Mechanical Engineering*, The War Office, 1940.
Henderson W.O., *Industrial Britain under the Regency: The Diaries of Escher, Bodmer, May and de Gallois, 1814–1818*, Frank Cass, 1968.
Hopkins E., *The rise of the Manufacturing Town: Birmingham and the Industrial Revolution*, Sutton Publishing, Revised Edition 1998.
Hoskins W.G., *About Britain No. 5, Chilterns to Black Country, Collins for the Festival of Britain Office*, 1951.
Hounsell D.A., *From the American System to Mass Production 1800–1932; The Development of Manufacturing Technology in the United States*, Johns Hopkins University Press, 1984.
Howard R.A., 'Interchangeable Parts Re-examined: The Private Sector of the American Arms Industry on the Eve of the Civil War', *Technology and Culture*, Vol.19, 1979, pp.633–649.
Howard R.A., 'Interchangeable parts revisited', *Technology and Culture*, Vol.21, 1980, pp.549–550.
Howe J.V., *The Modern Gunsmith*, Vols 1 and 2, Funk and Wagnalls, 1934, 1937 and 1941.
Jacques P., 'Peter Jacques inspects Enfield Inspectors Gauges', *Black Powder*, Vol.40, pp.26–29, 1993.
Jackson C.J., Ghosh S.K. and Johnson W., 'On the Evolution of Drill Bit Shapes', *Journal of Mechanical Working Technology*, Vol.18, 1989, pp.231–267.
Jackson M.H. and De Beer C., *Eighteenth Century Gunfounding*, David and Charles 1973.
Jervis J.W., *The rifle musket, A Practical Treatise on The Enfield Pritchett Rifle*, 1854.
Judd D., *Radical Joe, A life of Joseph Chamberlain*, University of Wales Press, 1993.
Kehaya S. and Poyer .J., *The Swedish Mauser Rifle*, North Cape Publications, 1999.
Kirkland, J.L., *How old French muskets were manufactured*, Libra Scientific, 1994.
Koenigsberger F., 'Production Engineering', in *History of Technology*, Vol.VII, ed T.I. Williams, pp.1,046–1,057, 1978.
Kolbe G., 'Making a rifled barrel', *Target Gun*, September 1991, pp. 41–46.
Labbett, P., British Service .577 inch Snider Ammunition, *Guns Review*, Vol.24, 553, 556–557, 1984.
Lancaster C., *The Art of Shooting*, 12th edition, 1954.
Landes D.S., *The Unbound Prometheus, Technological change and industrial development in Western Europe from 1750 to the present*, Cambridge University Press, 1969.
Langford, Dr, 'Birmingham Industries – Guns Rifles and Revolvers', *Iron*, 7 November 1874, XIV.
Lewis, J. ed, Greenwood & Batley Ltd, Centre for Historical Studies, Middlesex University. From the Ministry of Defence Pattern Room collection held in the Royal Armouries, Leeds.
Malone P.M., 'Little Kinks and Devices at Springfield Armory, 1892–1918', *Journal of the Society for Industrial Archeology, Theme Issue: Springfield Armory*, Vol.14, 59–76, 1988.
Moore, H., Beckinsdale, S. and Mallison, C.E., 'The Season Cracking of Brass and Other Copper Alloys', *Journal of the Institute of Metals*, Vol.32, pp.35–152, 1921.
Moore W.R., *The Foundations of Mechanical Accuracy*, The Moore Special Tool Company, 1970.
Mordecai A., *Military Commission to Europe in 1855 and 1856*, Washington, 1860.
Musson A.E. and Robinson E., 'The Origins of Engineering in Lancashire', *Journal of Economic History*, XX, pp.209-233, 1960.
Myatt F., *The Illustrated Encyclopedia of nineteenth century Firearms; An illustrated history of the development of the world's military firearms during the nineteenth century*, Salamander, 1979.
Newland M.A., *Gun and Gun Part Makers of Staffordshire*, Second Edition, pamphlet, 2000.
Nie D.A. and Bailey De Witt, 'William Tranter; Gunmaker to the trade', in *Gun Collectors Digest* ed Schroeder J.J, DBI Books, 3rd edition, 1981, pp.154–164.
Pam D., *The Royal Small Arms Factor,y Enfield and its Workers*, 1998.

REFERENCES

Parsons A.H., *History of the Institution of Mechanical Engineers*, Institution of Mechanical Engineers, 1947.
Pegler M., *Powder and Ball Small Arms*, The Crowood Press, 1998.
Porter M.E., *The Competitive Advantage of Nations*, Free Press, 1990.
Porter M.E., *Clusters and the New Economics of Competition*, Harvard Business Review, November–December 1998, pp.77–90.
Proof House, *James George Accles*, Typescript in the Birmingham Gun Barrel Proof House Collection, undated.
Prosser R.B., *Birmingham Inventors and Inventions, 1881*, Republished by SR Publishers, 1970.
Raber M.S., 'Conservative Innovators, Military Small Arms, and Industrial History at Springfield Armory, 1794–1918', *Journal of the Society for Industrial Archeology, Theme Issue: Springfield Armory*, Vol.14, pp.1–22, 1988.
Redford A., *The economic history of England 1760–1860*, Longmans, 1960.
Roads C.H., *The British Soldiers Firearm 1850–1864, from smooth bore to small bore*, Herbert Jenkins, 1964.
Roads C.H., 'The History of the Introduction of the Percussion Breech Loading rifle into British Military Service/The History of the British Military Rifle Musket 1850–1870', Ph.D thesis, Cambridge University, 1961.
Roberts N., *The muzzle-loading cap lock rifle*, 1952
Roe J.W., *English and American Tool Builders*, Yale University Press, 1916.
Roe J.W., 'Interchangeable Manufacture', *Proceedings of the Newcomen Society*, Vol.XVII, 1937, pp.165–174
Rolt L.T.C., *Tools for the job, a short history of machine tools*, Batsford, 1965.
Rosa J.G., *Colt Revolvers and the Tower of London*, Royal Armouries, 1988.
Rosenberg N., ed, *The American System of Manufactures, The Report of the Committee on the Machinery of the United States 1855 and the Special Reports of George Wallis and Joseph Whitworth*, Edinburgh University Press, 1969.
Rosenberg N., *Perspectives on Technology*, Cambridge University Press, 1976.
Saul S.B., 'The Market and Development of the Mechanical Engineering Industries in Britain, 1860–1914', *Economic History Review*: Second Series: XX, 111–130, 1967.
Schlesinger G., *Surface Finish*, Institution of Production Engineers, 1941.
Smith P.A., *Halls Military Breech Loaders*, Mowbray Press, 1996.
Shigley J.E., *Mechanical Engineering Design*, Second Edition, McGraw-Hill, 1972.
Smiles S., ed, *James Nasmyth Engineer, An Autobiography*, John Murray, 1883.
Smith B.M.D., 'The Galtons of Birmingham: Quaker Gun Merchants and Bankers, 1702-1831', *Business History*, Vol.XI, 1967, pp.132–150.
Smith D.M., 'Birmingham's Gun Quarter and its Workshops', *Journal of Industrial Archaeology*, Vols 1 and 2, 1964, pp.106–119.
Smith M.R., 'John H. Hall, Simeon North, and the Milling Machine: The nature of innovation among antebellum arms makers', *Technology and Culture*, 14, pp.573–591, 1973.
Smith M.R., 'The American Precision Museum', *Technology and Culture*, 15, pp.413–437, 1974.
Smith M.R., *Harpers Ferry Armoury and the New Technology; The challenge of change*, Cornell University Press, 1977.
Smithurst P., 'From handcraft to mechanised industry, developments in gun making in the nineteenth century', *Royal Armouries Yearbook, I*, 1996, pp.81–86.
Smithurst P., Glimpses into Greenwood & Batley, *Royal Armouries Yearbook, 3*, 1998, p131-136.
Smithurst P., 'The guns and gun making machinery of Robbins & Lawrence', *Royal Armouries Yearbook, 7*, 2002, pp.66–76.
Stelle J.P. and Harrison W.B., *A Gunsmith's Manual*, Excelsior, 1883, Reprint Gun Room Press, 1972.
Swenson G.W.P., *Pictorial History of the Rifle*, Ian Allan, 1971.
Tate D., *Birmingham Gunmakers*, Safari Press, 1997.
Taylerson A.W.R., Andrews R.A.N. and Frith J., *The Revolver 1818–1865*, Herbert Jenkins, 1968.
Taylerson A.W.R., *The Revolver 1865–1888*, Herbert Jenkins, 1966.
Taylerson A.W.R., *The Revolver 1888–1914*, Herbert Jenkins, 1970.
Taylor, L.B., *A Brief History of the Westley Richards Firm*, Shakespear Head Press, 1913.

Tholfsen, T.R., 'The Artisan and the Culture of Early Victorian Birmingham', *University of Birmingham Historical Journal*, Vol.3, No.2, 1952, pp.146–166.

Uselding P., 'Henry Burden and the Question of Anglo-American Technology Transfer in the Nineteenth Century', *Journal of Economic History*, Vol.30, 1970, pp.312–337.

Ure A., *Dictionary of Arts, Manufactures and Mines: containing a clear exposition of their principles and practice*, Longman, Orme, Browne, Green and Longman, 1839.

Walsh J.H., *The Modern Sportsman's Gun and Rifle*, Vol.2, 1884(?).

Walsingham, Lord and Payne-Gallawey R., *Shooting; Field and Covert*, Longmans, 1887.

Walter J., 'The Rise of the Piled Arms, A Short History of the Birmingham Small Arms Company', *Guns Review* 1984; **24**: pp.310–312, pp.397–399, pp.550–552, pp.610–611; 1986, **26**: pp.776–778, pp.826–827; 1987, **27**: pp.27–28.

Westley Richards, Catalogue, 2000.

Wilkinson H., *Engines of War*, 1841. Facsimile by Richmond: Richmond Publishing, 1973.

Williams D.J., (a), 'How did and do they make rifling', *Black Powder*, Vol.47, 2001, pp.82–86.

Williams D.J., and Johnson W., 'Schon's View of rifled infantry arms in the mid-nineteenth century', *Journal of Impact Engineering*, Vol.25, 2001, pp.315–330.

Williams D.J., (b), 'Getting it right the origins of precision manufacturing and mass production in gun making', *Black Powder Newsletter*, April 2001, 12–13, June 2001 p.25.

Williams D.J., 'A note on casting iron cannon balls: Ideality and Porosity', *Journal of Impact Engineering*, Vol.24, 2000, pp.429–433.

Williams D.J., '10 and 11 St Mary's Row: The Birmingham Gun Trade in Decline', *Black Powder Newsletter*, June 2002, pp.31–34.

Williams D.J., 'Whitworth and his gun gauges', *Arms and Armour*, Vol. 4, No. 2, 2007, pp. 110-121.

Williams D.J., 'James Farmer and Samuel Galton, the Reality of Gun Making for the Board of Ordnance in the Mid 18th Century', *Arms and Armour*, Vol. 7, No.2, 2010, pp. 119-141.

Williams, D.J., 'The Flintlock Ordnance Muskets of William III and their Supply', *Arms and Armour*, Vol. 9, No. 1, 2012, pp. 7-19.

Williams, D.J., 'Greener, William (1806-1869)', *Oxford Dictionary of National Biography*, Oxford University Press, Sept 2013.

Williams, D.J., 'Goodman, John Dent (1816-1900)', *Oxford Dictionary of National Biography*, Oxford University Press, Sept 2013.

Williams, D.J., 'Henry Nock, Walter Dick and Charles Lennox, The Duke of Richmond, and the journey to the final designs of the Nock screwless lock', *Arms and Armour*, Vol. 13, No. 2, 2016, pp. 28-47.

Williamson H.F., *Winchester, the Gun that won the West*, A.S. Barnes, 1972.

Winant L., *Early Percussion Firearms, A history of early percussion ignition – from Forsysth to Winchester, .44/40*, Spring Books, 1961.

Wise M.J., *On the Evolution of the Jewellery and Gun Quarters in Birmingham*, Transactions of the Institute of British Geographers, Vol.15, 1952, pp.59–72.

Whitworth J., *Miscellaneous Papers on Mechanical Subjects: Guns and Steel*; Longmans, Green, Reader and Dyer, 1873.

Whitworth Joseph and Co., Press Cuttings 1854–1862, Institution of Mechanical Engineers, Manuscript IMS 208.

Wood P.A., *Industrial Britain; The West Midlands*, David and Charles, 1976.

Woodbury R.S, *The Legend of Eli Whitney and Interchangeable Parts*, Technology and Culture, Vol. 1, 1959–60, p.23

Woodbury R.S., *History of the Milling Machine: A study in technical development*, The Technology Press MIT, 1960.

Woodbury R.S., 'Eli Whitney Armory Survey, Book Review', *Technology and Culture*, Vol.18, 1977, pp.146–148.

Woodworth J.V., *Drop Forging, Die Sinking and Machine Forming of Steel*, NW Henley Publishing, 1916.

INDEX

Adams & Deane factory   76
Accles, James   103-104
Allen, Ethan   60
American Civil War   46-48, 91, 158
American System of Manufactures   20, 55-72, 73-94
Ames Manufacturing Co.   84, 98
Anderson, John   58, 77, 79
Andrew, Jim   7, 147
Apprenticeship   21, 45-46, 63, 142
Archdale, James   20, 124
Armory Practice   55-72
Artifex and Optifex   7, 126, 140
Artisan   8, 19, 167,
Aston   31, 48, 125, 142
Automatic lathe   133

Babbage, Charles   45
Baker, Ezekiel   23
Barrel boring   32-33, 37-38, 46, 84, 147
Barrel grinding   31-33, 36
Barrel proving   42, 175
Barrel rifling   128-132
Barrel rolling   40-42
Barrel setting   80, 178
Barrel soldering   160
Barrel welding   8, 27, 32-34, 37, 40, 94
Battison, Edwin   69
Bayonet   12, 14, 40, 79, 103, 106

Bayonet makers   46
Bentley & Playfair   96, 98, 158, 170
Berdan rifle   103-04, 106, 180
Best gun   11, 17, 140, 162, 164, 168
Birmingham Gun Barrel Proof House   43-44
Birmingham Small Arms   96-103, 141
Birmingham Small Arms, founders   96-98
Black Country   18-19, 25, 31, 43, 57, 98
Blanchard lathe   57-59, 98
Blanchard, Thomas   58
Bodmer, Georg   22-28, 56
Boothroyd, Geoffrey   145-146
Bordesley   40
Boulton, Matthew   16, 18-19, 40, 54
Boxer cartridge   110-116
Box lock   137
Brades   38-40
Brace   24-25
Brazier   25-27, 80, 174
Breech loader   12, 17, ,65, 102-03, 106
Brown Bess   12, 21
Brown & Sharpe   68, 93, 132, 184
Buckland, Cyrus   58, 83
Bullet making machine   115, 121
Burton, James   66, 84, 106
Button rifling   131

Capital   69, 94
Capping breech loader   76, 95

Capping machines 115-116
Cartridge machinery 107-116, 121-22, 124
Carr Brothers 141-42
Carrington, James 65-67
Cherries 30-31
Cluster 10
Colt Armoury, Hartford 60, 73-75, 106
Colt, Samuel 73-77
Colts London Factory 76
Colvin, Fred 70
Contraction of the trade 49-51, 140-41, 145
Corporation Street 127
Craft 168-169
Crimean War 85, 91, 106
Cruxton, Malcolm, workshop 157-159

Damascus twist 32-36
Day man 52
Deane 76
Deep drawing 111, 114-15
De Witt Bailey 167, 175, 179
Depression 140
Deritend 31
Deyrup, Felicia 60
Dickens, Charles 76
Domestic system 52, 169
Draw knife 25, 52
Drilling jig 74-76, 117, 135
Dudley Castle 18
Dunham, Keith 145

East India Co. 23, 27, 78-79
Enfield Manufactory 73-90
Enfield rifle, Pattern 1853 12-16, 75, 85-93, 103
Engineering 9, 139
Engraving 146, 162, 164
Eskilstuna Rifle Factory 133

Factor 51
Factory system 46, 50-54, 167
First World War 45, 97-99, 131, 134, 139-141
Filing 21, 24-25,
Fitting 21-25
Flintlock 12, 21, 28, 56, 174-176
Ford, William 141-142
Forging, forgings 27-29, 38-40, 56, 75-76
Form cutter 62

Galton, Samuel 43
Gatling Arms and Ammunition Co. 103
Gauges 22-23, 67-68, 84-90, 99-100, 169, 171
Georgian houses 52, 142
Goodman, John 93, 95-98, 179
Gordon, Robert 60, 71
Great Exhibition 73, 77
Greener, William 32 et seq.
Greener, William Wellington 23, 48, 85, 103,
--104, 126-27, 137
Greenwood, Thomas 106-116

Greenwood & Batley 106
Gunmakers Arms 146

Hall, John 60, 64-67
Hall rifle 65
Handsworth 9, 19
Harpers Ferry Armoury 64-66
Hartford 60, 73-75, 103
Hill, Haydn, workshop 162-63
Hook cutter 130
Horse 153

India Pattern Brown Bess 12-13, 79, 162
Industrial Revolution 18-19
Inletting 53-54
Inletting or recessing machine 80, 82, 92, 97,
 109-10, 185
Institution of Civil Engineers 31, 68, 73-76, 77
Institution of Mechanical Engineers 69, 107-13, 177
Interchangeable manufacture 11, 55-72, 73-94,
 167-69
Interchangeablity of Colt's revolvers 75
Internal contracting 68-70

Jig borer 134-135
Johansson gauge blocks 133-35
Johansson, Carl 134
Jones, John 28-29, 42, 85

Kynoch 121-25
Kyte, Robert, workshop 147-50

Lee Enfield 12-13
Lee Metford 12
Lewis Guns 99, 124
Lewisham 27, 84, 126
Liège 22, 126, 140
Limit gauges 68, 134, 177
Lock making 24-31
Lock nomenclature 16
London Armoury Co. 91, 103
Lovell, George 56, 75, 83-85, 93, 129, 131
Lumps, machining 147, 151, 152, 162, 163
Lunar Society 19, 43

McFarland, Cary 98-99
McKinley tariff 140
Machine tools 56-57, 95 et seq., 137 et seq.
Manby, Charles 73
Manchester 95-96,
Manufacturing miller 60, 63-64
Martini Henry rifle 12-13, 17, 103
Maudslay, Henry 77
Mauser rifles 105, 135-36
Micrometer 49, 132, 135, 179
Middletown 60
Milling machine 7, 10, 31, 59, 60-65, 67, 69,
 76, 83, 92, 116, 121, 126, 132, 139, 147,
 ---162, 168

INDEX

Movements in the trade   49-52
Mordecai, Alfred   70, 92
Muntz, George Frederick   78
Napoleonic Wars   45-46, 56, 79, 84
Nasmyth, James   28, 31, 77
National Arms and Ammunition Co.   103
New Buildings   101, 158
Newdigate, Sir Richard   21
Nock, Henry   56
North, Simeon   60

Oldbury   38
Onyans, William   94
Ordnance System   22-23
Organisation of manufacture   69
Osborne, Henry   40
Outworkers   52

Parker Hale   160
Pattern   22, 91
Pattern Room   119
Percussion lock   16
Piece man   52
Polhem sticks   134-35
Portsmouth block making machinery   58
Practical interchangeability   11, 67
Pratt & Whitney   62, 135
Price Street   142, 146, 153, 158, 175
Production volumes   43, 128, 140-01, 175
Prosser, Richard and Richard Bissel   85, 129

Raglan, Lord   78
Receiving gauge   67-68, 84-91
Remington   68, 70, 77, 94, 127, 131
Report of the Select Committee on Small Arms   78-83
Revolver manufactuirng   117-125
Rifling guide   128 et seq.
Rifling machine   129, 131
Robbins & Lawrence   7, 60-63, 66, 68, 73, 77,   ---83, 91, 92, 107, 124
Roe, Joseph   11, 69
Rolling   32, 40-43, 94
Root, Elisha   77, 178
Round Oak Ironworks   139
Rosenberg, Nathan   176
Ryder forging engine   76, 133
Royal Manufactory at Enfield, see Enfield Manufactory

St Mary's   50-51, 141-45, 158-61
Season cracking   115
Second World War   131, 140, 141
Scientific Management   68
Screw mill   29
Sharps Rifle Co.   61
Sine bar rifling machine   131
Skelp   40
Slave gun   43-45

Slave trade   43-45
Slotter   139, 152
Smethick   31, 32
Smith, D.M.   50, 175, 181
SMLE   98
Smoke lamp   25, 121
Snider conversion   12-13, 102-03
Socket bayonet   12, 14
Specialisation   21, 65
Spencer, Christopher   132
Sporting gun making   137 et seq.
Spring vice   26
Springfield Armory   8, 48, 60, 63, 67, 71, 77
Springfield musket   71, 77
Stamps   27-28, 38, 40, 178
Starr, Nathan   63, 68
Steam engine   31 et seq.
Stork   46, 96
Swage rifling   131

Tipton   18
Tolerances   11-12, 49, 60, 92, 133
Tower   22-23
Tower, Bagot Street   23, 104-05
Toy trade   19
Trade gun, see slave gun
Tranter   124
Treadle lathe   138, 150-51
Tula Armory   27, 29, 65, 85, 107
Tumbler mill   29-31
Turn screw   24, 25

Universal milling machine   132

Vernier Calliper   93

Wallis, George   77
Warner, Thomas   60, 67, 77
Weaman Street   51, 117, 124
Webley   117-125
West Bromwich   31
Westley Richards   8, 95-96, 139, 162, 168-169, 180
Westley Richards Arms and Ammunition Co.   103
Whitney Armory   64- 65
Whitney, Eli   59-60
Whitney, Miller   69
Whitworth, Joseph   77, 79-81, 95, 130 132
Wilkinson, John   25, 40
Winchester   68, 70, 127
Wise, M.J.   50, 175
Woodbury, R.S.   62, 69, 176
Wood screw making   179-80
Woolwich Arsenal   77-79
Workshops   142, 147-163

Other titles published by The History Press

A Birmingham Backstreet Boyhood
GRAHAM V. TWIST

All the harshness of daily life is remembered here by local author Graham Twist. Despite hard living conditions and a distinct lack of money, a strong community spirit prevailed and families and neighbourhoods were close-knit. The womenfolk in particular took great pride in their homes, however humble, and scrubbed their front steps and swept the areas in front of their houses religiously.
978 0 7509 4965 1

Birmingham Pubs
KEITH TURNER

A century ago there were more than 1,500 hotels, inns, taverns and public houses in Birmingham. This book, using more than 200 photographs and other illustrations from the late 1850s to the early 1970s, forms an evocative record of some of them. City centre watering-holes, back street boozers and village inns are recorded here and this book serves as a unique record of an important part of Birmingham's social history.
978 0 7524 1809 4

Birmingham Voices
COMPILED BY LUCY HARLAND

The people who share their memories and photographs in this book all took part in the BBC's Millennium Oral History Project, *The Century Speaks*. They were chosen because of their range of ages and backgrounds would present a broad picture of local experience, opinion and knowledge. What results is an important and interesting collection that ranges from the humorous to the intensely personal and moving.

978 0 7524 1848 3

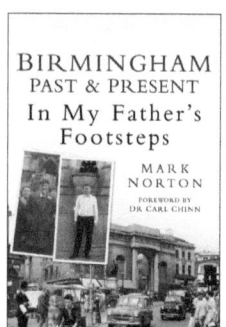

Birmingham Past & Present
MARK NORTON

In the 1950s and '60s, Dennis Norton took his camera and went to work photographing buildings in areas of the city due for redevelopment, capturing a Birmingham that is now gone. Almost half a century later Mark Norton discovered these photographs, taken by the father he never knew: Dennis died just nine weeks before his son was born.
978 0 7509 4504 2

Visit our website and discover thousands of other History Press books.
www.thehistorypress.co.uk